GUIDE TO SITE SURVEYING

edited by Ralph Hewitt

ARCHITECTURAL PRESS · LONDON

ISBN 0 85139 268 7

Photoset and printed in Great Britain by
BAS Printers Ltd, Wallop, Hampshire

Contents

Acknowledgements

The editor is indebted to John Ransford BSc (Eng) FICE MBCS, Kenneth Fresson AA Dip RIBA, Leslie Fairweather RIBA, Peter Burberry Dip Arch RIBA and Alan Hodgkinson M.Eng C.Eng FICE Struct. for their valuable revisions and advice.

Technical data about the instruments and equipment referred to in this book was supplied by the following:

A. H. Hall & Bros. Ltd.
142 Upper Street, London N.1
44 Barkston House, Croydon Street, Holbeck, Leeds 11
16 Ellesmere Road South, Chorlton Cum Hardy, Manchester M21 1TE
25 Somers Road, Rugby

Francis Barker & Son Ltd., Fircroft Way, Edenbridge, Kent

Hilger & Watts Ltd., 98 St. Pancras Way, Camden Road, London NW1

Holmes Bros. (London) Ltd., Stirling Works, Simonds Road, London E10

The M.D.S. Ltd., 41 St. James's Gardens, Holland Park, London W11 4RG

Survey & General Instrument Co., Fircroft Way, Edenbridge, Kent

W.F. Stanley & Co. Ltd., 20–22 Mortimer Street, London W1

Vickers Instruments Ltd., Haxby Road, York

Wild Heerbrugg U.K. Ltd., Revenge Road, Lordswood, Chatham, Kent.

Rabone Chesterman Ltd., Hockley Abbey Works, Birmingham 18

The photograph on the cover was taken at Cubitt's Thamesmead site and the theodolite was loaned by A. H. Hall & Bros. Ltd., Islington.

Recent developments in Surveying

The two basic problems in surveying an area of land are the measurement of length and angle. During the past thirty years a number of extremely sophisticated methods of making these measurements have been proposed, some of them researched, and a few of them marketed. Many surveyors have been slow in accepting this new equipment, possibly because of respect paid to long-established methods of surveying, but more probably because of the financial investment represented by stocks of traditional instruments and the high cost of the newly-developed instruments. However, as we are at present learning to work with metric units, there is a unique opportunity for innovation in surveying practices which may never be repeated.

The surveyor now needs an increasingly intimate mixture of the old skills and the new technology. The most striking recent advance in surveying has been the growing use of non-physical distance measuring techniques. This has made large area survey by tri-lateration (the determination of the relative positions of the corners of a triangle by measuring the lengths of all three sides) more convenient than by tri-angulation.

It is sometimes said that a good surveyor is one who can carry out a survey without a chain or tape. The various items described in the course of this book lend credence to this saying since with this equipment a great many survey operations may be carried out without making direct linear measurements. The instruments may be divided into two main categories, electronic and optical.

Electronic instruments

Two Electronic Distance Measuring instruments (EDMs) have become well established in most parts of the western world. They are the geodimeter and the tellurometer.

Geodimeter

The geodimeter, invented in 1949, uses a visible light signal. Light from a common incandescent lamp or from a mercury arc lamp is modulated in intensity at a known frequency. The modulation provides the measurement data. A geodimeter set consists of an active transmitter-receiver unit and a passive reflector. **1** and **2**. The light beam is directed from the transmitter to the reflector and back to the receiver. **3**. When the complete set of

1 *AGA Geodimeter Model 8*

2 *Zeiss EDM Reflectors*

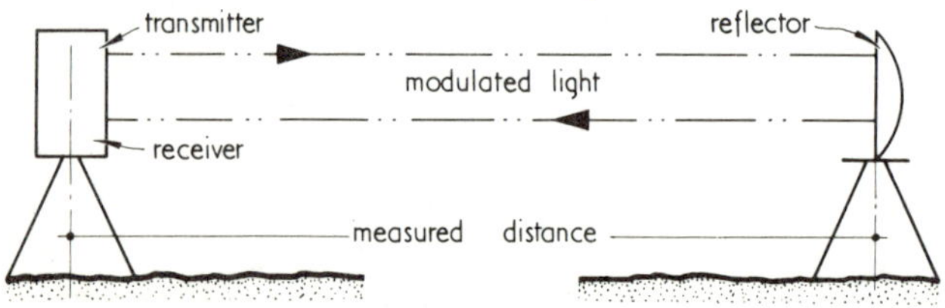

3 *The Geodimeter*

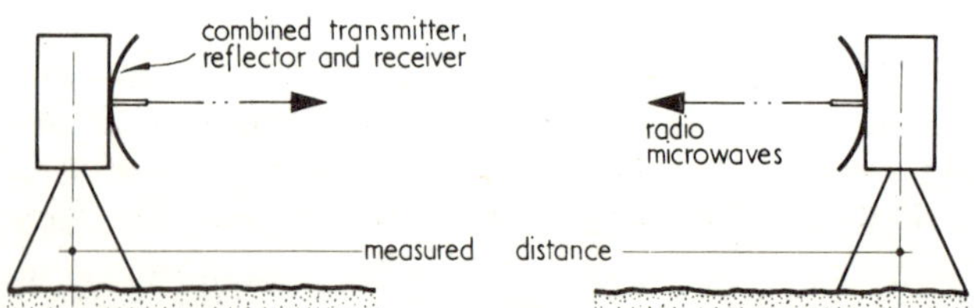

4 *The Tellurometer*

observations onto a station has been made the surveyor has to solve three simultaneous equations to find the distance. Accuracy of 10mm + 2/1000 000 of the distance can be achieved without special precautions. For some surveys this will be unnecessarily high but the convenience of measuring long or short lines quickly, regardless of the intermediate ground conditions is a great advantage. The remote reflector is generally of the retro-directive prism type. This is simply a corner cut off a cube silvered on the outside. It has the great advantage that it will accept light from a wide range of directions and reflect it along the same path. Before this development plane mirrors were used but they needed too precise and stable alignment for general field use. A single prism reflector need only point to within 10° of the transmitter, but for longer sights up to nine reflecting elements are used which must then all be in the same plane and facing the transmitter so that each prism is at the same distance and the modulation of the reflected signal is not blurred.

Tellurometer

The tellurometer was invented in 1957. A measuring set is comprised of two active electronic units, each consisting of transmitter-receiver circuitry. In measuring a distance each unit is placed over one end point and transmission of radio microwaves is directed towards the other. **4**. The carrier and modulation frequencies of one unit differ from those of the other by a fixed amount. An advantage of the tellurometer is that the signal can penetrate haze, mist, smoke and even light rain. This instrument, to a much greater extent than the geodimeter, requires skilled operatives at both ends of the survey line. The check procedure for fault finding in setting up and in operation is daunting eg. 'With a new tube EF 95 replace tubes V 101 and V 105 one at a time. If still not normal replace V 102, V 103 and V 104 with new EF 91 one at a time'. The tellurometer's working accuracy is similar to that of the geodimeter. Specialised models of the tellurometer are made, one for use in aircraft and the other for hydrographic work. The 'Aerodist' can be used for line ranging and enables lines up to 300km to be measured. It operates with an airborne master and two remotes set up at fixed points. The 'Hydrodist' tellurometer replaces the sextant angle method of determining the position of survey craft within distances of up to 40km of the coast. Two masters located in the craft take simultaneous readings onto two remotes set at fixed points on a shore.

Other electronic equipment

Later additions to the range of EDMs are the Zeiss electronic recording tacheometer **5**, and electro optical telemeter **6**, and the Wild D110 and D150 Distomats **7** and **8**. Other newer EDMs are in the development and production stage but they do not promise performances significantly different from their predecessors (see chapter 8 Table VIII). The most obvious difference between one EDM and another is in the type of electro-magnetic signal that is transmitted. These include visible light, infra red light, radio signals and laser beams. Each has its adherents, its advantages and, no doubt, its disadvantages.

For work of geodetic standards, where the curvature of the earth would prevent direct line-of-sight observation, extensive use has recently been made of radar and instruments carried on aircraft. This method has lately been augmented and extended by observing onto artificial satellites, whose orbits are known to a high degree of accuracy.

Optical instruments

A wide range of optical distance measuring instruments has been developed and most are fairly well known. Range finders may have an accuracy

5 *Zeiss Reg Elta 14 Tacheometer*

7 *Wild D1 10 Distomat*

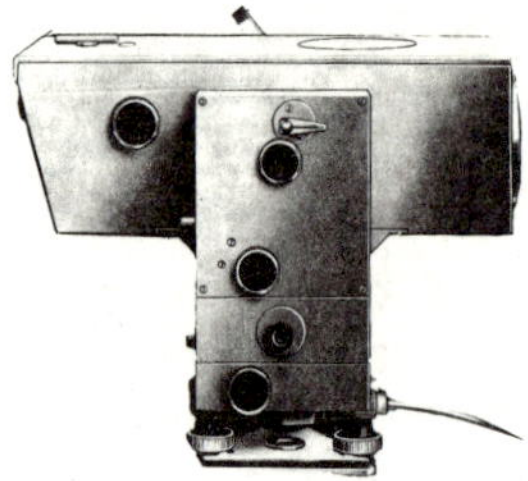

6 *Zeiss EOK 2000 Electro-optical Telemeter*

8 *Wild D1 50 Distomat*

of up to 1 in 1600 and are used for the rapid location of minor detail. They have been found particularly useful where heavy traffic would make direct measurement impossible.

Tacheometry

Stadia and tangent tacheometry are the traditional methods of optical distance measurement, but these methods have suffered because of the time and tedium involved in reducing observations (reduction of observations is necessary when a quantity is not measured directly, but is deduced from other measurements). The increasing use of desk computers capable of simple programming will probably invigorate this method of survey which is quite suitable for minor detailing and finding spot heights for contouring. With the addition of wedge attachments accuracy is increased from 1:1000 to 1:5000 or 1:15 000 and precise traversing for engineering works is possible.

Vertical staff tacheometers with self reducing systems have an accuracy of about 1:1000. These instruments are very popular as they combine speed in operation with ease in recording the results. **9** and **10**. Greater accuracy is possible with the horizontal bar tacheometer which again has the advantage of automatic reduction.

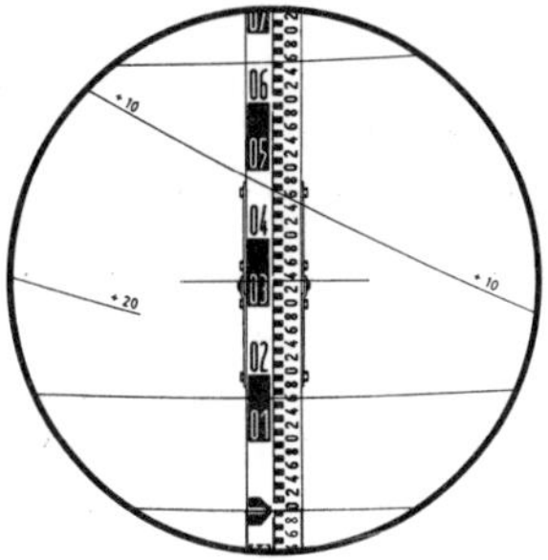

9 *Zeiss RTa4 Tacheometer and Field of View*

11 *Watts Subtense Bar*

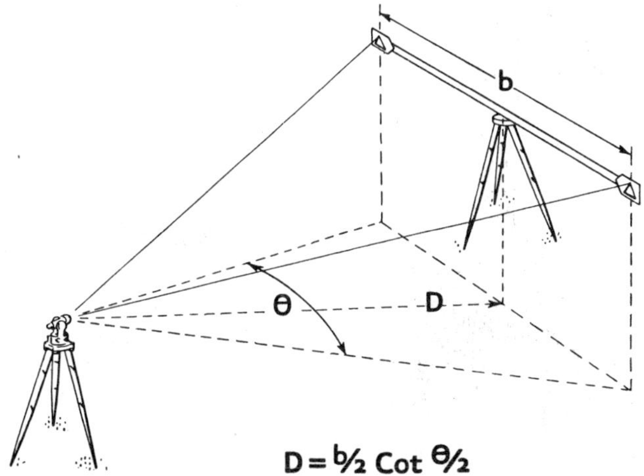

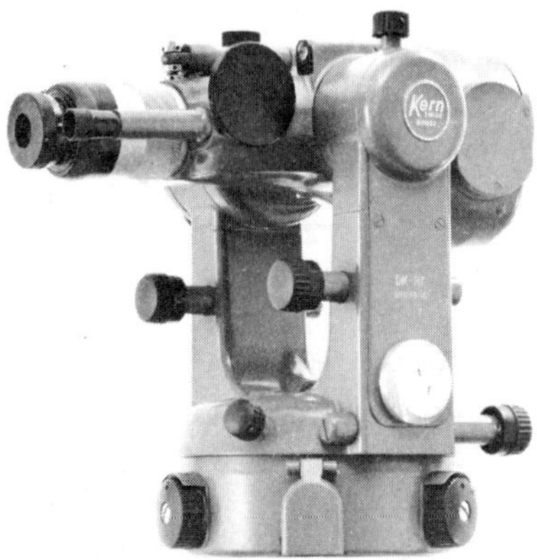

10 *Kern DK RT Tacheometer for Horizontal Rod*

12 *The Reduction of Subtense bar readings (Courtesy of Watts)*

Subtense bar

A less commonly used instrument is the subtense bar. The greater time taken in using this is compensated for by the greater accuracies possible. The use of the subtense bar is very suitable for precise traversing and in the location of ground control points for air photography. Reduction of observations is easily done with a small book of tables. **11** and **12**. Accuracy decreases with distance from 1 in 28 000 at 15m to 1 in 1300 at 300m.

Plane table surveying has been modernised by the use of self reducing telescopic alidades that have an accuracy similar to that of tacheometers. **13** and **14**.

Levels and theodolites

Surveying has also advanced in the increasing precision possible from the level and the theodolite. An example of the precision level is the Wild N3 which may be used for first and second order levelling. It incorporates a parallel plate micrometer in front of the objective lens and with it height differences of 0·01mm may be estimated. **15**. The astronomical theodolite is an instrument of astonishing accuracy. **16** illustrates the Zeiss 002 that has an accuracy of 0·3 seconds of arc. This is the angle that would be subtended at the eye by a man standing one thousand kilometers away i.e. from Lands End to John O'Groats.

A growing range of applications for surveying instruments and techniques exists. Some idea of the diversity of possible uses may be gathered from the following. The Meteorological Office uses a specialised form of theodolite for measurements of the Upper Winds by means of pilot balloons. Because of the extreme angles of elevation involved in observations onto high floating balloons, the eye piece of the telescope is let into the side of the instrument and it remains horizontal for all positions of the objective. **17**.

Photography

An interesting offshoot of terrestrial photography has been the development of the stereometric camera. **18**. This has found uses in forensic photography, photography for the preservation of cultural relics, architecture, archeology and civil and mechanical engineering projects. It has proved invaluable where the accurate planning of three dimensional assemblies is required i.e. the manufacture of automobiles, ships and aeroplanes. More bizarre applications have come in the fields of anthropology, ethnology, medicine and zoology where the subject of study has been too brittle,

13 *Watts Microptic Alidade*

14 *Wild RK1 Self Reducing Alidade*

15 *Wild N3 Precision Level*

16 *Carl Zeiss Jenner Theodolite Theo 002 Angles measured to 0·3 sec.*

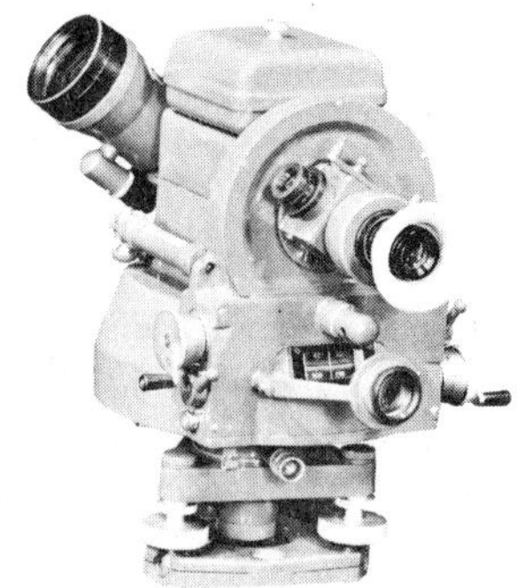

17 *Hilger and Watts SMI pilot balloon Theodolite*

18 *Wild C12 Stereometric Camera*

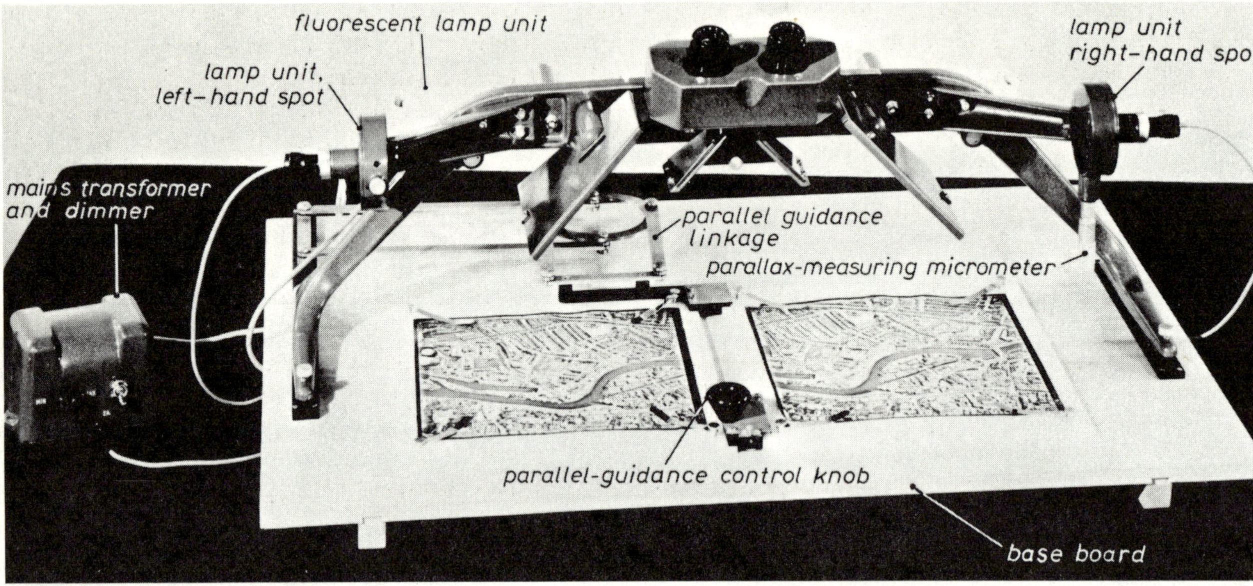

19 *Hilger and Watts SB 190 Scanning Stereoscope*

confused, ill or hungry to permit long sessions with tape measure and callipers.

The basic achievements of aerial photography are now widely known i.e. the rapid and accurate mapping of extensive areas with only a few accurately placed control points needed.

Less well known is the wealth of extra information that skilled readers are now taking from aerial photographs. Photo interpretation is most often done using a scanning stereoscope. **19**. It has been found an advantage to train outside specialists to use the stereoscope rather than hope to teach a stereoscopic operator the widely different and complex subjects relevant to an investigation.

Forestry experts have quickly learnt to distinguish different types of tree and the presence of spring growth by slight tonal differences in a photograph. This has been made easier by the use of various coloured filters in front of the camera. Quantitative measurements from photographs have resulted in the accurate estimation of the amount of timber in an area of forest and of animals in a large herd. A great deal more information will obviously be available with the promised arrival of colour photographs. A surprising fact is the possibility of detecting disease in crops by the tonal change in a photograph; disease has often been detected in this way before the farmer on the ground had suspected its presence. Films are available that react to colours outside the visible spectrum, and using these the photo interpreter gathers information that simply is not there for the unaided eye. Geologists have found the aerial photograph invaluable in tracing fault lines, synclines and anticlines etc. Photographic evidence has helped in finding minerals, water, oil and natural gas for the geologist. It has also shown a wealth of historic and prehistoric settlement areas of interest to the historian and anthropologist; there is in some areas an embarrassment of riches. Geologists find that they can obtain more information from photographs taken higher up and they welcomed the availability of high definition photographs taken from satellites orbiting at about 300km.

Satellite transmission of continuous television pictures has made it possible to chart the movement of icebergs and hurricanes. It should also be possible to detect schools of whales and shoals of fish, although unless conservation measures are taken this information would quickly lead to over fishing. It may have seemed that the advent of aerial photography, EDMs, ODMs and the computer would result in a lot of redundant surveyors. The reverse is in fact true. The surveyor is now in a position to offer many new services in a greatly increased market, and the future looks even more interesting and varied.

Part 1
Surveying instruments

Introduction

This book deals with the surveying of sites and with the setting out of buildings. The surveying of existing buildings is not covered. Surveying in the present context is the location of points three-dimensionally. This means fixing positions vertically and in two dimensions on the horizontal plane.

Site surveying

The architect wants to know the following facts:

(a) shape and dimensions of the site on plan;

(b) levels of the different parts of the site and the relation to these to adjacent roads, services and so on;

(c) location of any immovable objects on the site, their size on plan and elevation

(d) location of points of entry to the site

(e) run of water courses

(f) location, lines depth and size of services

Setting out

The setting out of buildings has been described as surveying in reverse. Starting with a sufficiently accurate survey of the site and with the working drawings, the architect has to determine the following:

1 location of the foundations to all walls

2 the line of all services and the position of manholes etc.

3 location and outline of all site works and the gradient of roads and paths

Instruments

Before a survey can be undertaken the surveyor must be fully aware of the equipment that is available to him and how he should use it. He will also need to make a realistic assessment of the accuracy needed in the survey. If the accuracy achieved is insufficient then the survey may be useless, or at least difficult to work from. An over accurate survey should be avoided since it must cost more in time and money than the merely adequate survey. To help appreciation of the existence of a range of accuracies the instruments described in Part 1 have generally been introduced in ascending order of accuracy.

Chapter 1

Surveying instruments: Classification by function

This chapter outlines the basic operations involved in surveying work and relates them to types of instrument classified according to function, eg instruments for establishing vertical orientation, for measuring angles, and so on. Subsequent chapters deal with each functional group in turn

General

Instruments may be grouped according to function and graded within these groups in ascending order of precision. Classification by function can be somewhat arbitrary because many instruments are capable of fulfilling more than one function. For this reason it is convenient to consider them according to exclusive functional features of their design. Seven groups result, and in chapter two instruments are arranged as far as possible in order of ascending precision in use. The functional groups are:

Instrument groups

Vertical orientation
Linear measurement
Establishing right angles
The determination of azimuth bearings from true or magnetic north
Measurement of differences of height
Measurement of horizontal angles
Measurement of vertical angles
Measurement from photographs

Vertical orientation

Most surveying instruments must be orientated vertically before use. This means setting them up so that they are located exactly over a reference point on the ground. In surveying, the terms nadir and zenith are often used to describe the hypothetical extremes of a truly vertical line passing through the reference point and the earth's centre of gravity.

Survey measurements in plan, ie the horizontal plane, must relate to this nadir/zenith line.

The vertical orientation of surveying instruments is normally made by means of plumb bobs and lines or optical plummets.

The vertical orientation of points above ground level, especially when tall buildings are concerned, is usually made with special forms of optical plummet or by means of optical attachments to other instruments. (See Chapter 2.1)

Linear measurement

Measurements are made from one reference point to another, variously referred to as station points, object points or target points, the first being the most usual term.

Physical measurement is made with chains, flexible tapes or rods marked out in units of length with various subdivisions.

Indirect linear measurement can be made and this is done by optical means described in Chapter 2.2.

Indirect linear measurement can also be made with surveying instruments equipped to measure horizontal or vertical angles, or with special instruments such as tacheometers or rangefinders. (See Chapter 2.2)

Right angles

The relationship of curved or irregular boundaries, or the relative location of extra points to a line (or its extension) between two station points can be determined by measurement at right angles from it.

There is a number of instruments available for the quick determination of right angles along a given line, including optical squares and cross-heads. These are described in Chapter 2.3.

Orientation

In navigation, in large scale surveying and whenever permanent reference is important it is necessary to determine true north. This may be done by solar or astronomical observations or worked out from magnetic north with the help of an almanack or Ordnance Survey data.

Magnetic north is often sufficient orientation for architectural surveying, provided that the date is given on the relevant drawings. It may be observed either by careful reference to Ordnance Survey or other survey maps, or with a magnetic compass.

Magnetic north is observed by means of a magnetised needle supported by a pivot point at its centre of gravity. Instruments differ in methods of supporting needles, protection against damage, size, delicacy, and refinements such as means for damping down oscillation. (See Chapter 2.4)

Height

Direct measurement of height is made either by means of tapes, or a graduated measure referred to a horizontal plane, ie levelling.

The normal instruments used in surveying for levelling work are optical levels—dumpy, quickset, precise etc, which are designed to establish a horizontal plane of observation—and a measuring

rod or staff.
The measuring staff is held on various points and sighted through the instrument so that differences of height between the line of sight and station reference point can be observed. (See Chapter 2.5)

Horizontal angles

The plan relationship of two station points to a third or other points can be established either by right-angular measurement or by distance and angular measurement.
The conventional manner of representation and measurement is to define points two-dimensionally on a horizontal plane (plan) and note differences in height between them.
When angular measurements are used it is possible to establish the locations of all but two primary points by calculation without resort to linear measurement. (See Chapters 10 and 13)
Accuracy in plan location demands that angular measurements shall be truly horizontal. Various instruments are available for the purpose including levels fitted with a horizontal circle, plane tables and, especially, theodolites. (See Chapter 2.6)

Vertical angles

An alternative to direct measurement of height difference, once plan location has been determined is by calculation and involves measuring vertical angles. (See Chapter 2.7 and Chapter 6)
Calculation makes use of angles, measured relative to the horizontal plane, from one station point to another. It is essential that such angles shall be measured in a truly vertical plane and the normal methods employ optical instruments capable of accurate reference to the horizontal plane.
Instruments available for measuring vertical angles are pocket levels such as the abney, levels fitted with gradienters (to find gradients), telescopic alidades (plane table) or, especially, theodolites. (See Chapter 2.7)

Chapter 2·1

Surveying instruments: Vertical orientation

Vertical orientation of almost all surveying instruments is essential for their proper and accurate use. In the case of levels, theodolites and similar instruments the measurements they make refer to a point on the ground, which must lie on the vertical axis of the instrument. Again angular and height measurements relate to a truly horizontal plane, and the only way to establish one is to set the instrument concerned so that it measures at right angles to a truly vertical axis ie by reference to gravity
This section describes instruments and attachments for enabling vertical orientation to be established

Plumb lines and bobs

The simplest means of establishing vertical orientation, (ie locating the position of a point on the ground vertically beneath a reference point) is by means of a plumb-bob and line **1**. This is essentially a weight suspended in line with its axis of gravity. Plumb-bobs are most frequently used in conjunction with optical surveying instruments to establish the location of their vertical axis on the ground. They are suitable for only comparatively rough vertical orientation.

Spirit levels

More sophisticated methods depend upon some means of reference to the earth's gravitational field, usually a spirit level; sometimes a pendulum and occasionally both. When these are used it becomes necessary to relate vertical axes to a horizontal plane. This is most often done by means of spirit levels, of which the centre bubble type is used to obtain a close approximation.

Circular spirit levels

The essential features of centre bubble spirit levels are illustrated in **2**.

1 *Plumb-bob and line*

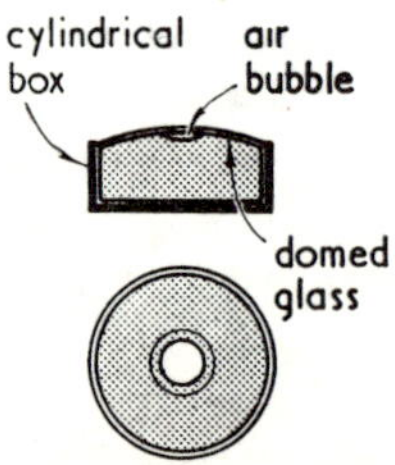

2 *Section and plan of centre bubble or arcular spirit level*

The circular bubble sometimes serves as a convenient substitute for a plumb-bob—when attached, for example, to the back of a levelling staff on a right-angled bracket. It has the advantage that it centres both north-south and east-west at the same time. It is normally used on surveying instruments to obtain a first approximation when setting them up.

Tubular spirit levels

For final accurate adjustment of non-automatic instruments a tubular spirit level is necessary. This is usually built-in and it consists of a closed glass tube delicately ground internally along its length to the arc of a circle and partially filled with alcohol, ether or some other low viscosity liquid. **3**, **4**.

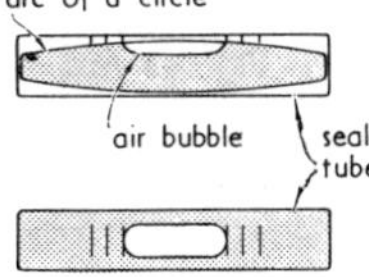

3 *Section and plan of tubular level*

4 *A typical tubular level*

The sensitivity of a level increases with the radius of the arc to which it is ground and is usually expressed as the angle through which the tube must be tilted to cause the bubble to travel a certain distance eg 8min per 2mm. Bubble-length varies with temperature and so graduations are usually marked on the tube at 1mm (or tenths of an inch) intervals on both sides of the bubble at its central position.

When used to determine vertical axes tube spirit levels have to be used in pairs or successively in two positions at right angles in plan **5**. A single tubular level only establishes verticality in one direction eg North-South or East-West.

Spirit levels are built into almost all optical instruments used in surveying. In the more accurate there are optical arrangements which display the bubble through an eyepiece so that it appears as two halves sliding past each other during adjustment. Coincidence of both halves indicates the bubble is central. This method improves the accuracy of setting by a factor of eight **6**.

Optical plummets

Where plumb lines would not be sufficiently accurate for vertical orientation, especially where considerable heights are involved, optical plum-

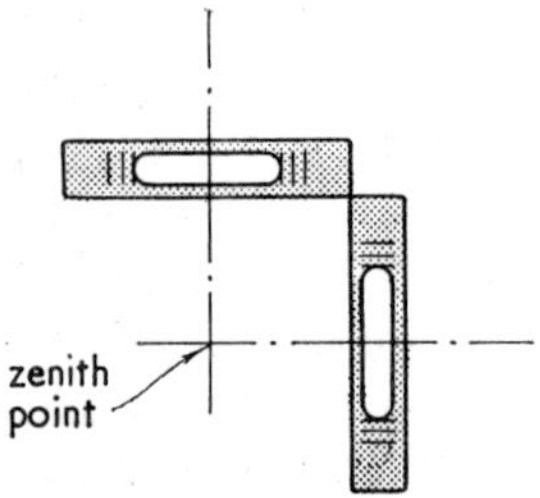

5 *Spirit levels placed at right angles to determine vertical axis*

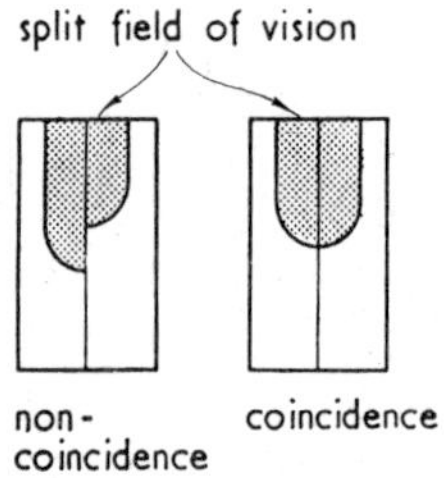

6 *Coincidence bubble as it appears in instrument*

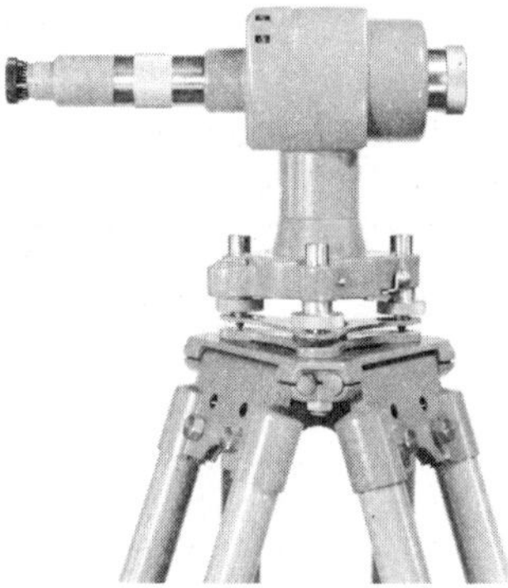

7 *Wild ZBL1 centring plummet, a type of optical plummet fitted to the tripod of an instrument such as a theodolite to establish vertical orientation accurately*

8 *Wild ZBL optical plumb, a type of optical plummet which reads up as well as down*

mets are used. Many instrument makers supply them as an extra for levels or, more often, theodolites, but it is becoming more usual to find them built-in.

Where extreme accuracy is required special optical plummets can be obtained. These are separate instruments screwed to the tripod of a theodolite or level etc to establish vertical orientation before the theodolite is fitted **7**, **8**.

The essential features of an optical plummet are shown in **9**. These instruments are sometimes made so that they can extend a vertical axis upwards to the zenith point as well as downwards.

The increase in high building construction has made it expedient to manufacture instruments especially designed for vertical orientation. The optical plummets illustrated in **7**, **8**, **10** and **11** all incorporate telescopes to extend their range. This is for use when, for example, readings have to be taken to the bottom of an excavation or to the top of a tall building. They call for a precision difficult to obtain with instruments that rely upon spirit levels to establish horizontally, and several now incorporate semi-automatic levelling devices **10**, **11** making use of a pendulum principle which is explained under 'automatic levels' in Chapter 7.

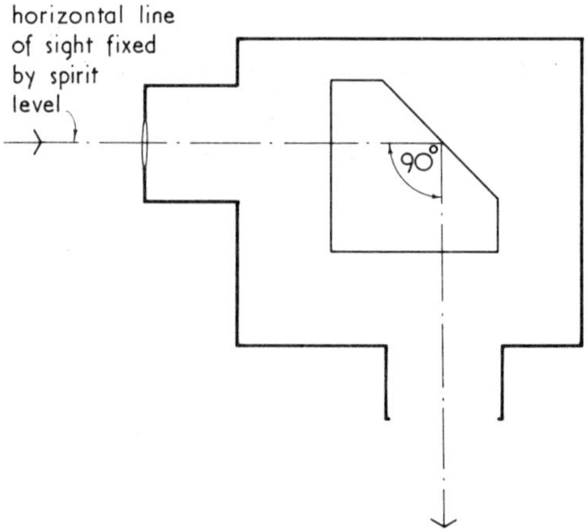

9 *General principle of an optical plummet*

10 *Hilger & Watts Autoplumb, a type of optical plummet (sighting upwards only) which sets itself in vertical orientation automatically*

11 *Hilger & Watts Autoplummet, similar to 10 but sights downwards only*

Chapter 2·2
Surveying instruments: Linear measurement

This section describes instruments ranging from tapes to special optical devices which are intended especially for linear measurement in surveying. It includes some notes on special provisions made in some other instruments, such as levels, for the same purpose

LINEAR MEASUREMENT

Ranging rods

These are wood or metal poles painted in 0·5m deep bands of contrasting colour. They are used to mark station or intermediate points on lines of measurement (ranging lines) and to measure short offset lengths approximately **1**, **2**.

Folding measuring rods

These are made from wood laths in short lengths, jointed to open out to 2m in length, and marked in dm, cm and 2mm. They are used to measure offset lengths, heights and short linear distances **3**, **4**.

Land chains

Chains are made from 8wg to 12wg steel or galvanised iron wire in links which, joint-to-joint, constitute the units of measurement. Chains are made 100 links long and marked every tenth or fifth link with a tally. Gunters chains (traditional) are 66ft long; engineers' chains are 100ft. Metric chains are usually 20 metres with a tally every metre. Chains are used to measure comparatively long linear distances **5**. Heavy brass swivel handles are fitted to the ends of the chain and are included in the measurement.

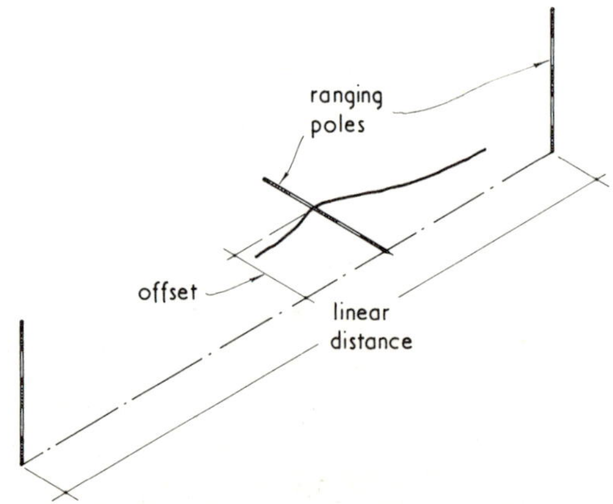

1 *Ranging rods used to mark out surveying line*

2 *Holbro ranging pole laid out as offset rod*

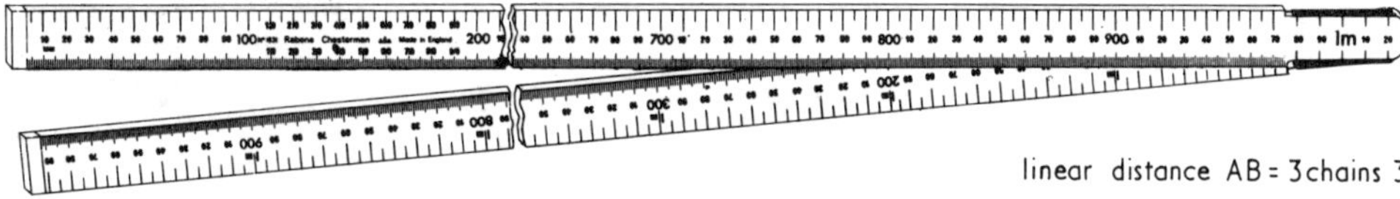

3 *Rabone Chesterman two-fold rod*

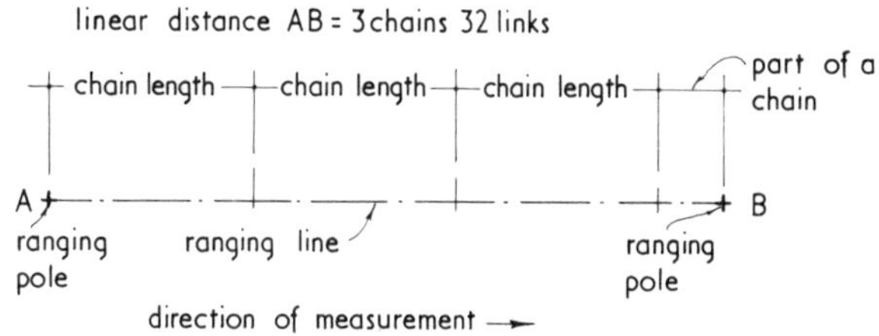

5 *Chain measurement along a survey ranging line*

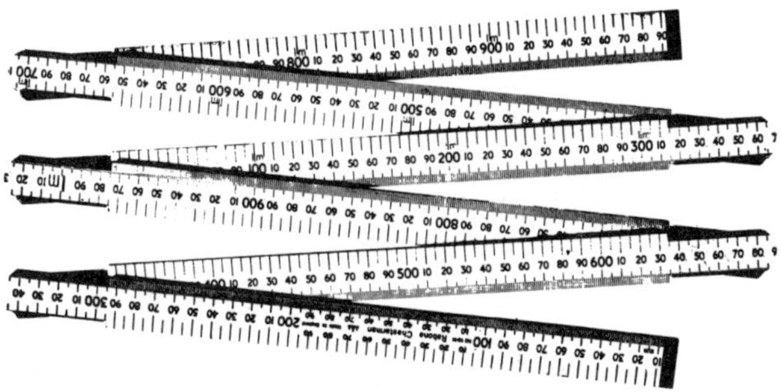

4 *Rabone Chesterman multi-fold rod*

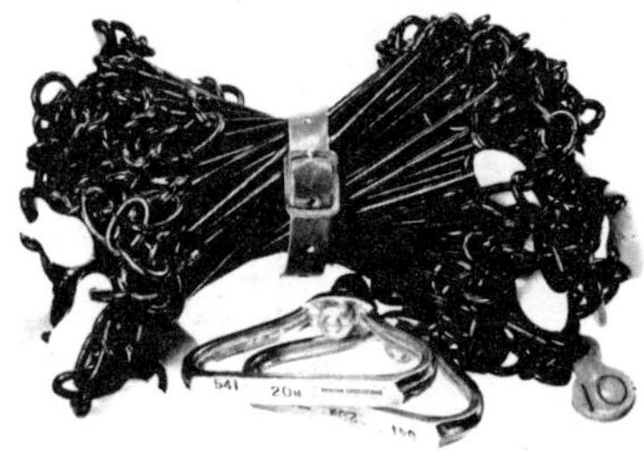

6 *Rabone Chesterman land chain. Yellow tallies at every metre, red numbered tallies at every 5 metres from end to centre to end. Tallied in accordance with BS 4484: 1969*

Steel band chains

These are made from narrow 3 or 5mm wide, blued steel tape marked with brass studs in units similar to land chains **7** but the end lengths are usually further subdivided. They are used instead of land chains for great accuracy and ease of movement.

Etched steel bands

These are made from 10mm wide carbon steel in 50m and 100m lengths and are used where even greater accuracy is required **8**. They are rather more delicate to handle than ordinary steel band chains.

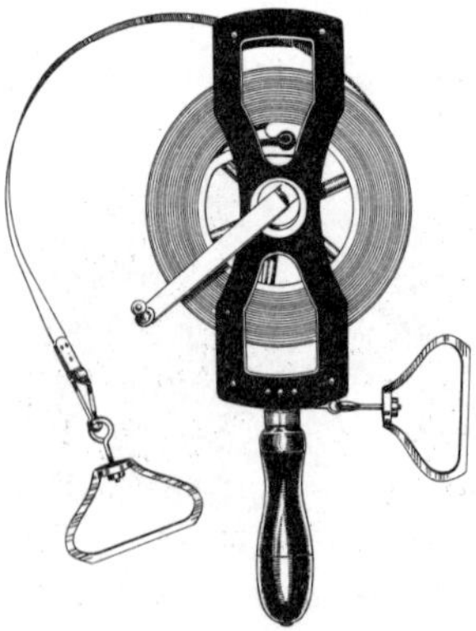

7 *Rabone Chesterman steel band chain*

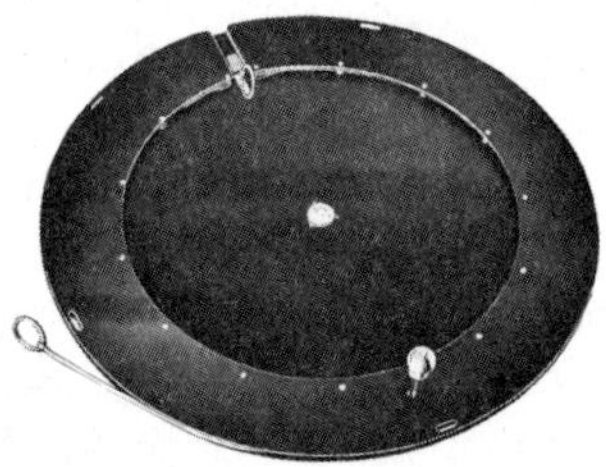

8 *Rabone Chesterman etched steel band drum*

9 *Rabone Chesterman metallic and non-metallic tape*

Measuring tapes

These are made from 15mm wide plastic tape, either plain or reinforced with copper wire (metallic) **9**.

Graduations

Tapes may be marked, on one side only, in metres, cm and 5mm, or else with feet, inches and quarter inches on the reverse side. They are used instead of chains or for long offsets in land survey or for buildings. They are more accurate than land chains, but less accurate than steel bands.

Steel tapes

Steel tapes are made from 12mm wide carbon steel with markings etched or enamelled on. They are marked as are metallic tapes but down to millimetres and eighths, and used where a high degree of accuracy is needed **10**.

Standard invar bands

These are made from 3mm or 5mm wide invar tapes, artificially aged and all in one piece. This metal is chosen because of its high degree of dimensional stability due to its low coefficient of expansion (0·000 0009 per °c).

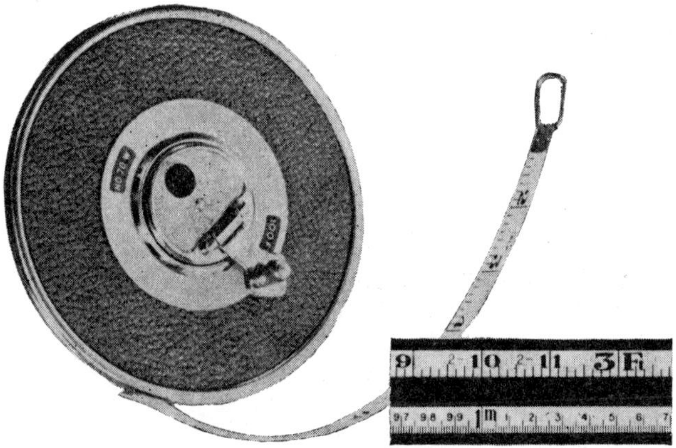

10 *Rabone Chesterman silver line steel tape with enlarged section of the tape*

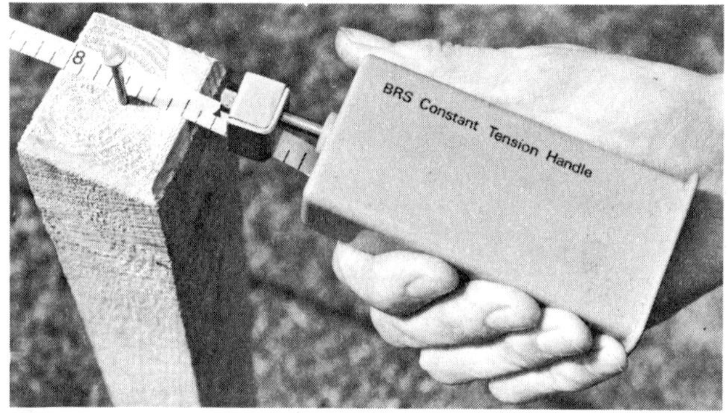

11 *BRS Constant tension handle*

Graduations

Invar bands are marked with fine lines and, at the terminals, 5mm either side of the zero mark is divided into millimetres and fifths of a millimetre. Lengths may be graduated 50m or 100m at intervals of 5m, 10m or 20m.

Standards of measurement

Invar bands are used as standard for other linear measuring instruments or where a very high degree of accuracy is needed for base lines etc. If necessary invar bands may be submitted to the National Physics Laboratory for tests in lineal accuracy, coefficient of expansion etc.

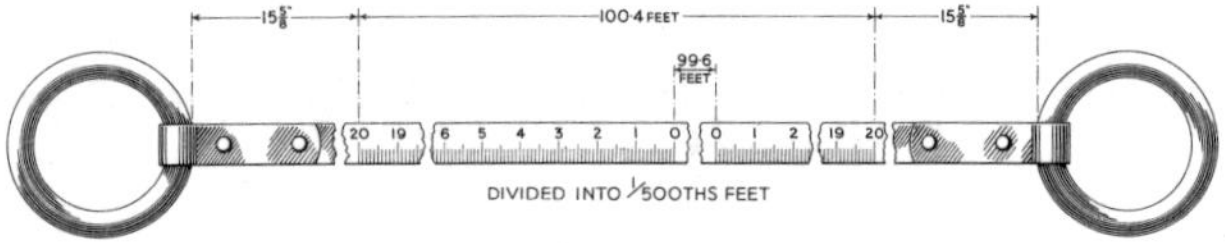

12 *Rabone Chesterman invar band. A metricated version is now available.*

Rangefinders

Rangefinders are optical instruments with a pair of mirrors or prisms set accurately a fixed distance apart and capable of rotation against a scale graduated to record distance ($\frac{1}{2}$ base Tan variable angle) as and when two images of the 'object' coincide in a central eyepiece. They usually measure distances starting about 100m away and up to 10 000m. They can be graduated in metres, yards, chains etc, as required. The principle is illustrated in **13**. A rangefinder theodolite is shown in **14**.

Tacheometers

A tacheometer or telemeter is also an instrument for determining distances without physically measuring them, but operates on a different principle from the rangefinder. The principle, known as the stadia system of linear horizontal measurement is illustrated in **15**.

Instrument distance C usually takes the form of two hair lines equidistant from the horizontal centre line of the reticle seen in telescopes of levels,

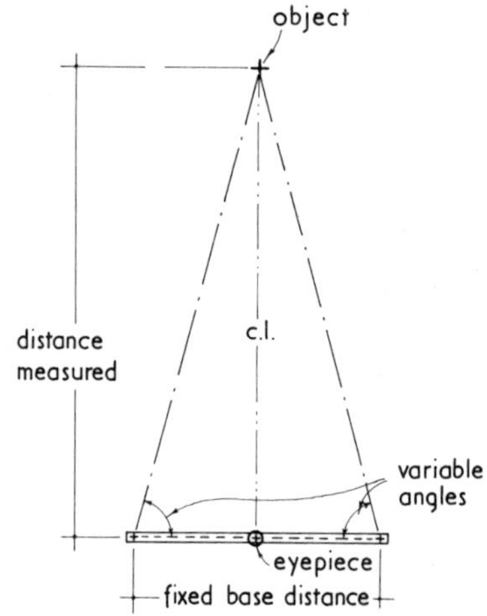

13 *The principle of measuring distance with a rangefinder*

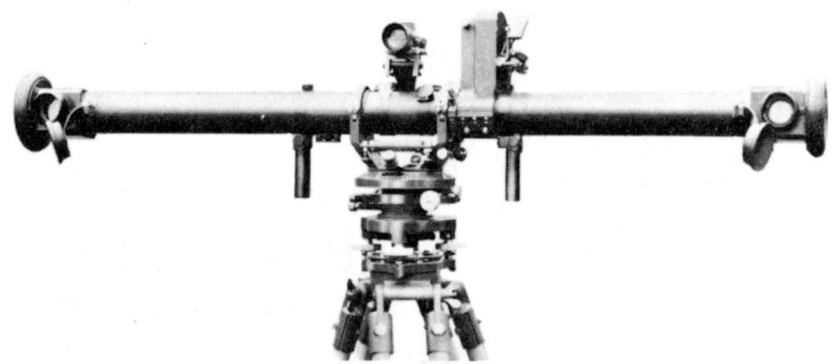

14 *Hilger & Watts CS 10·1 rangefinder theodolite*

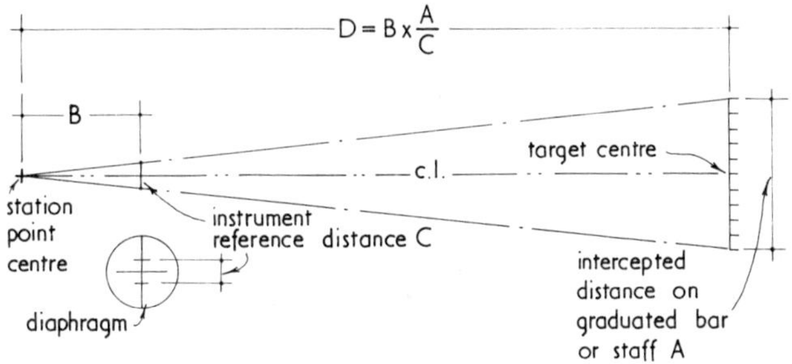

15 *The principle of the stadia system of linear measurement horizontally*

theodolites etc. The distance B is more complex than shown since it involves the optical system of the telescope but nevertheless may be regarded as a fixed dimension in modern instruments. Dimensions B and C are *within* the instrument and represent a ratio called the stadia ratio. The appropriate figure for an instrument, usually 100, is given by the manufacturer. Distance D is then 100 times the reading A on the graduated bar or staff at the object point.

The simplest way of using instruments for making stadia readings is to have the line of sight horizontal and the staff vertical. This is not always possible on steep slopes. It becomes unnecessary when a horizontal subtense bar is used with a suitable instrument instead of a telescopic staff read with a horizontal line of sight. This method is described in detail in Chapter 4.

Stadia measurements on slopes

On slopes a situation like that shown in **16** arises: three stages of calculation become (theoretically) necessary. First, to determine the vertical equivalent (a_2b_2) of the distance intercepted (a_1b_1) at the object point. Second, to find the actual distance from station to object (D) and third, to work out the horizontal or plan distance (D_2), all depending upon the angle of elevation in depression Θ.

The process is called 'reduction' and several methods and devices have been introduced to cut down the labour involved: these include tacheometric tables and slide rules. Modern practice favours the Beaman stadia arc, which is also fitted to many older instruments, and direct-reading instruments generally called 'self-reducing tacheometers'. For a description of this technique see Chapter 5.

The tangential system

A second method of determining distances optically (ie without physically measuring them) is by the tangential system for which an instrument called a gradienter is required. This is described below.

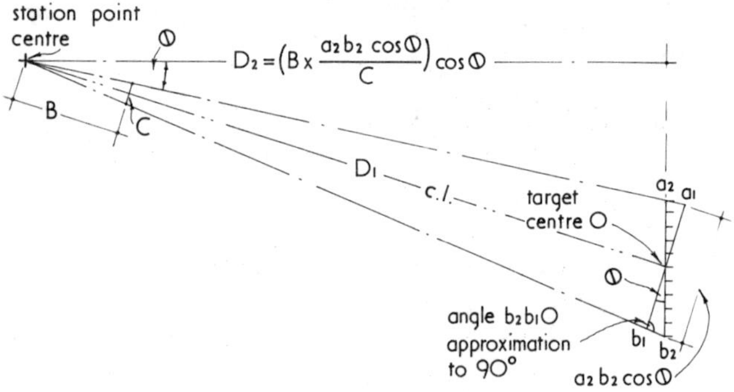

16 *Stadia measurement on slopes*

The simplest case of tangential measurement, from which the system derives its name, is illustrated in **17** where all that is necessary is to read the vertical angle O and the staff intercept (a) and apply the formula: D (horizontal distance) = a Cot Θ.

Gradienters

Many instruments, both levels and tacheometers, are fitted with a 'gradienter' **18**. This is a knurled tangent screw which is usually fitted at the eyepiece end of the telescope. It generally has a micrometer head graduated so that each turn moves the line of sight through a height of 1m at a distance of 100m, although other dimensions may be used. It follows that after setting the telescope level and taking a reading on the staff (object) the screw may be turned and a second reading taken, such that 100 × No of turns × reading difference (a) = horizontal distance D.

Tangential measurement on slopes

When it is impossible to sight the staff horizontally the calculation of D involves much more work **19**. This can be reduced if angles whose natural tangents are 0·01, 0·02 etc are set on the instrument to read from the horizontal position when the intercept (a) can be read off a graduated staff at the object point relative to these angles. If two successive angles from a table of such angles are used their tangent difference is always 0·01 and the calculation of distance and height can be made mentally **20**. It is possible to use instruments on which predetermined angles are set.

Special tacheometers

The predetermined angle method has not been used much in the past because a very high quality instrument is necessary. However there are now special instruments whereon direct readings of tangent values is possible, employing the principle

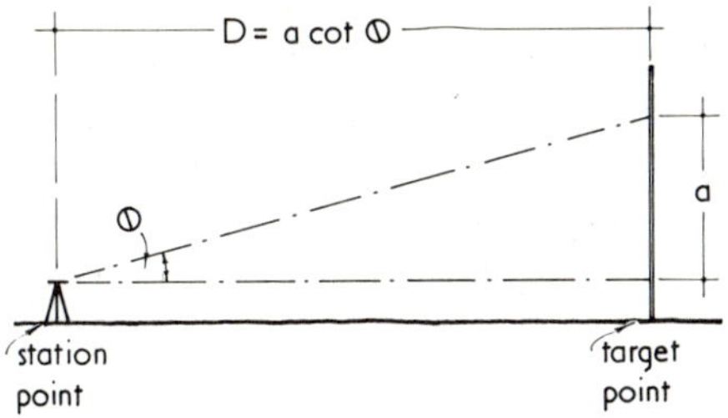

17 *Linear measurement by the tangential system with a horizontal line of sight*

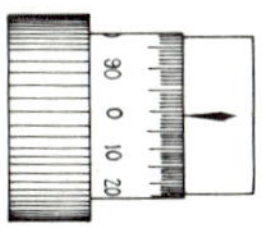

18 *Graduations on the micrometer head of a gradienter*

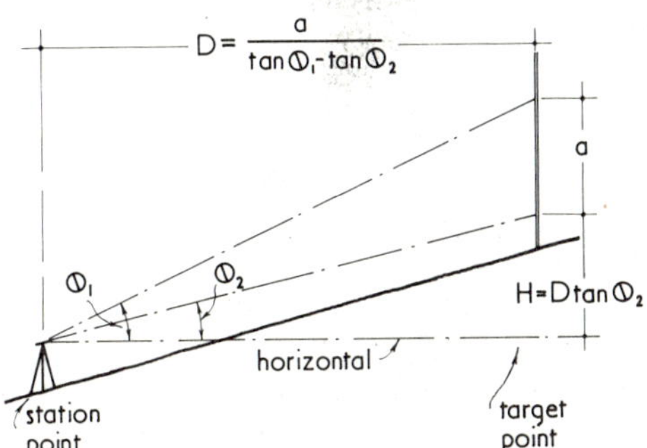

19 *Linear measurement by the tangential system on a slope*

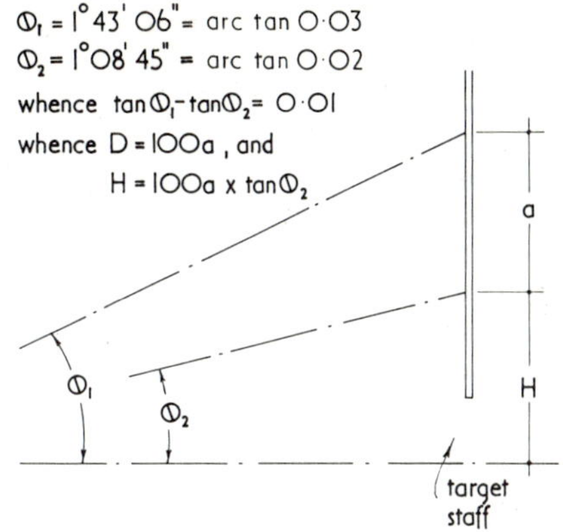

20 *Tacheometrical principle of measurement with a tacheometer working by displacement through a predetermined angle*

given above. A typical recent and labour-saving type of instrument is the double-image tacheometer. This is an instrument that makes use of deflection of rays through a prism placed in front of the object glass of its telescope in such a way that displacement on a graduated target staff equal to 0·01 of the distance from station point to target can be read off directly **21**. It still remains to reduce D, as shown in **19** to horizontal distance but even more sophisticated instruments are available which give a reading of the necessary correction as well. These are called double-image-self-reducing tacheometers **23**, **25**. The accuracy obtainable with

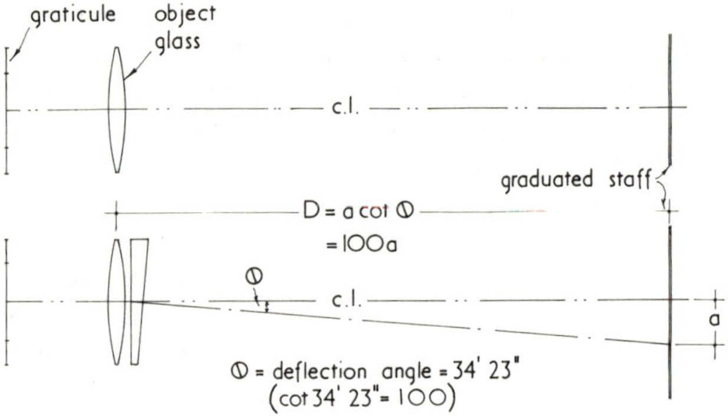

21 *Angular displacement with a variable prism*

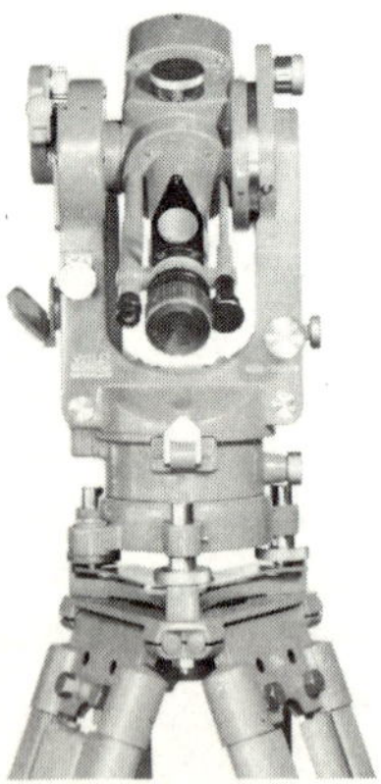

23 *Wild RDH tacheometer*

24 *Wild DM 1 telemeter, an attachment for fitting to a theodolite*

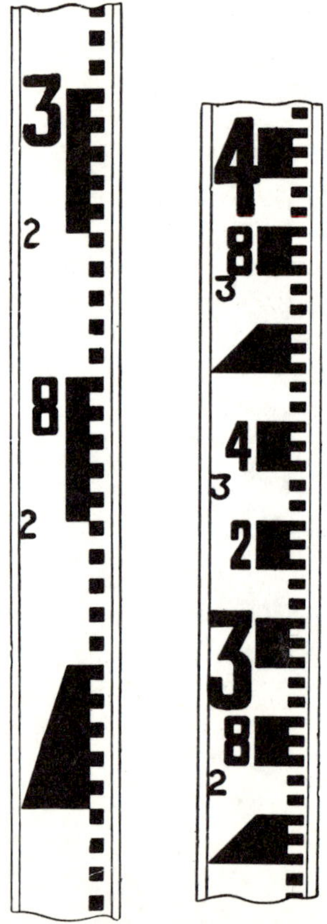

22 *Graduations on two typical tacheometric staves*

25 *Wild RDS tacheometer*

due care exceeds that of measurement by taping. Alternatively attachments can be obtained for fitting to theodolites **24**.

Tacheometric staves

The graduated staff or subtense bar used at the target point in tacheometry is, more often than not, an ordinary levelling staff (described elsewhere) but special staves and bars are desirable for long sightings and are necessary for really accurate work. **22** illustrates some typical patterns of staff graduation.

Subtense bars

Invar subtense bars are used for accurate work on the tangential system **30** using a theodolite (described in Chapter 2.7 but employing a fixed horizontal distance at the target point **26**. This has the advantage that it is independent of elevation since the horizontal angle remains constant despite it. Arrangements for illuminated points at the graduation marks are usually made and tables giving distances (in metres or feet) corresponding with observed angles are supplied.

Double image subtense bars

Double-image tacheometers require a special form of subtense bar or 'horizontal staff'. These have a complete graduation across the bottom edge and three vernier graduations, one central and one at each end. They are mounted horizontally on a graduated vertical rod so that the bar can be set at the same height as the instrument and thus give a direct height difference **27**. One of the vernier graduations appears, through the action of the prism in the instrument, to be located as shown in **28**. A proprietary type of double-image subtense bar is shown in **29**.

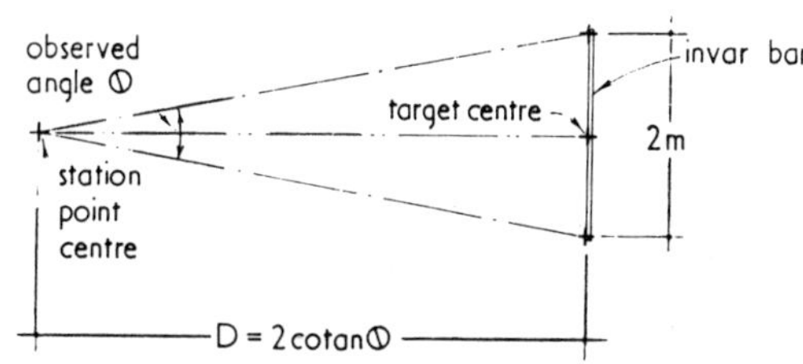

26 *Linear measurement using subtense bar*

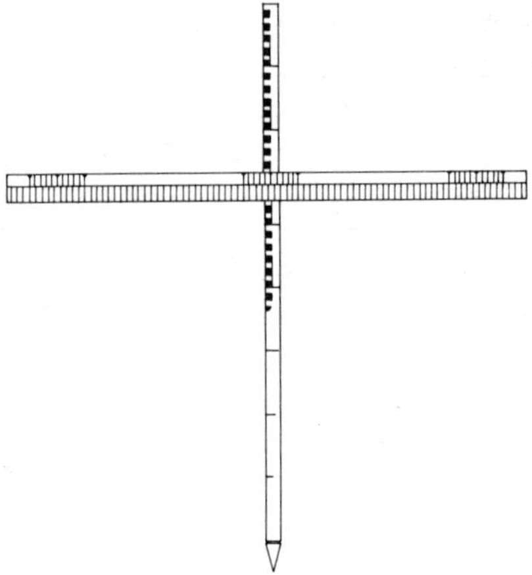

27 *a double-image subtense bar and supporting graduated rod*

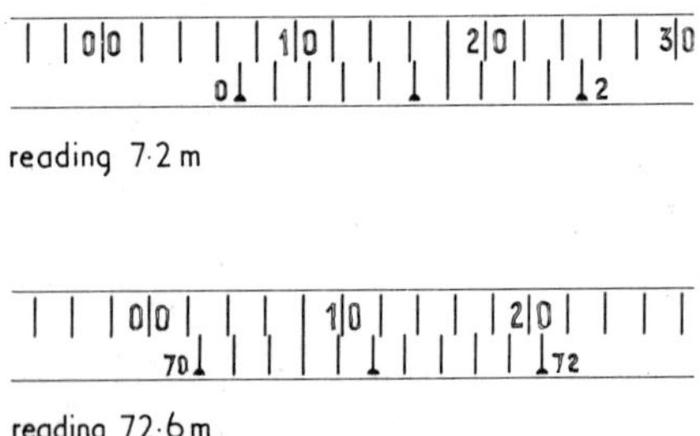

28 *Displacement verniers on double-image subtense bars as seen through the tacheometer*

29 *Wild telemetric staff* (*a double-image subtense bar*)

30 *The Watts 2-metre subtense bar has targets supported on invar-steel rods which have so low a coefficient of expansion that distance measurements are unaffected by wide variations in temperature*

Chapter 2·3
Surveying instruments: Right angular measurement

This section describes instruments and techniques for establishing right angles during surveying operations

General
The need to set out right angles from a chain line to measure offsets or establish points in a grid can be met with various degrees of accuracy. The 3: 4: 5 triangle **1** is well known, but simple instruments such as those described below are often more convenient.

Ranging rod sighting piece
One very useful instrument is a special ranging rod which incorporates a metal cylinder having four slits in it set so that right angles can be sighted through **2**.

Sighting squares
Another method of right angular measurement is by means of instruments known as squares or cross-heads which are fitted to the top of a rod. Examples are the open cross-head, **3**, and the French cross-

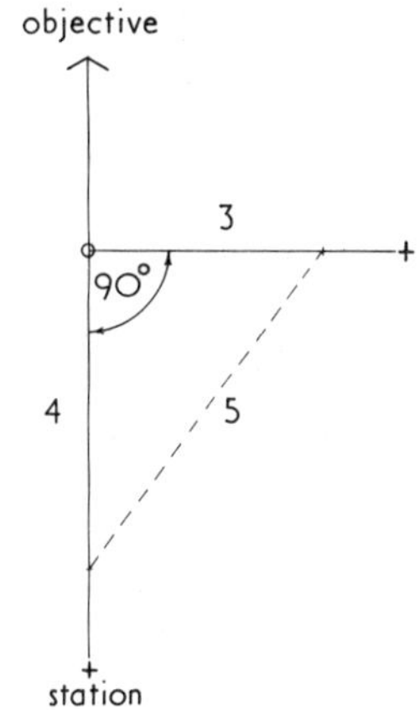

1 *The 345 right angle*

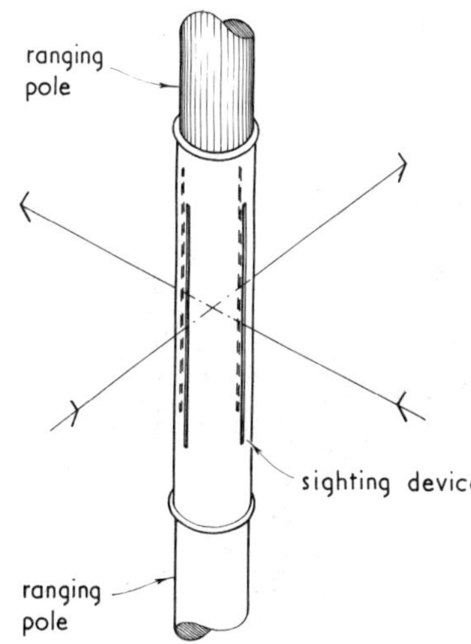

2 *Right angle sighting slits built into a ranging rod*

head, **4** which both make use of sighting slits similar to **2**.

Optical squares

More accurate and more convenient determination of right angles is possible with optical squares. Two kinds are normally available: mirror and prismatic. Mirror optical squares are flat cylindrical boxes with an arrangement of mirrors as shown in **5**. They have the disadvantage that they may need adjustment; they are also less robust than prismatic squares, of which the principle is illustrated in **6**.

A proprietary type is shown in **7**.

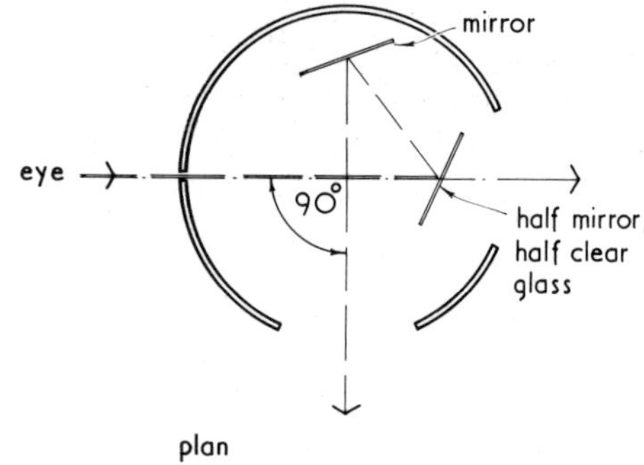

5 *The optical principle of a mirror square*

3 *Stanley open cross-head*

4 *Stanley French cross-head*

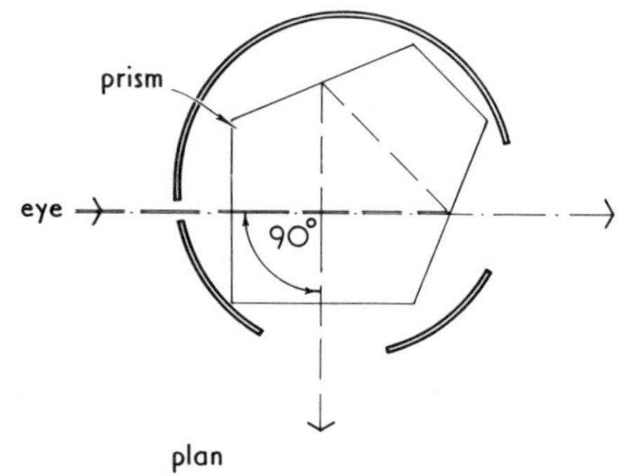

6 *The optical principle of a prismatic optical square*

7 *Francis Barker optical square*

Chapter 2·4

Surveying instruments: Azimuth orientation

This section describes some of the types of compasses used in surveying for orientation to magnetic north, or for taking azimuth bearings

MAGNETIC COMPASSES

Trough compasses

Magnetic north should always be observed, however roughly, during a survey. One of the simplest instruments in use for the purpose is the trough compass **1**, **2**. This is a narrow rectangular box in which a relatively long magnetic needle is supported upon an agate point. It may be set upon a plane table (see Chapter 6) and a line drawn against the side of the box indicates magnetic north with sufficient accuracy for most purposes. More refined and accurate versions of the trough compass are made for attachment to theodolites.

Prismatic compasses

Prismatic pocket compasses are probably the best-known type since they are widely used for military purposes. They have a magnetic needle attached to a circular card graduated down to half-degrees. This can be read through a prism eye piece at the same time as an object is sighted through an object vane, thus giving its magnetic bearing from the observer's position **3**.

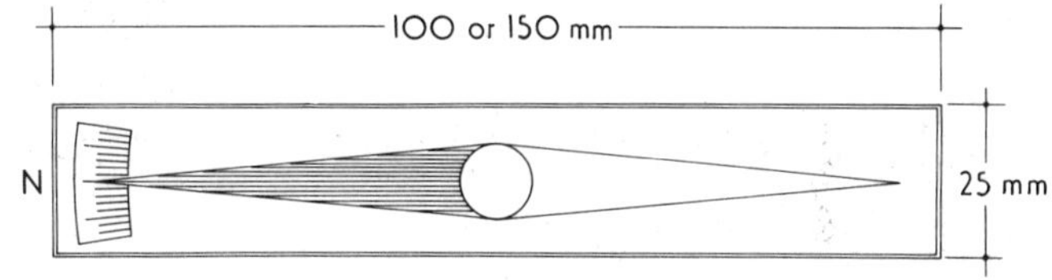

1 *Plan view of typical trough compass*

2 *Francis Barker trough compass*

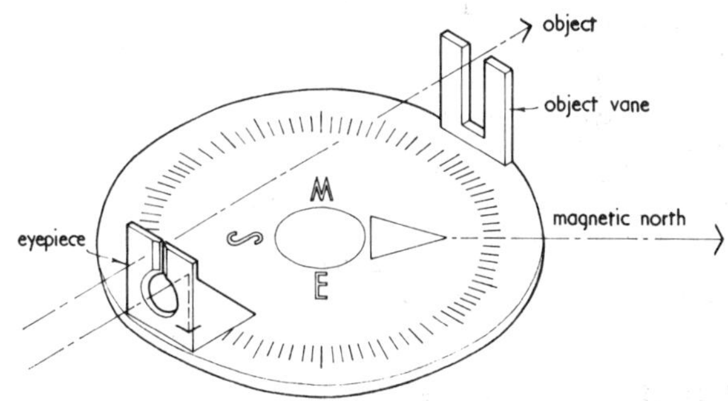

3 *General principle of prismatic compass in perspective*

Types of prismatic compass

Two types of prismatic compass are normally available. One kind has a jewelled pivot which can be lifted so that the card is free to rotate **4**, the other has the needle and card immersed in liquid to reduce the effects of friction and shock **5**.

Instrumental compasses

More sophisticated larger compasses are made for more accurate work. These are constructed either with sighting devices (or telescopes) analogous to the prismatic compass, or with a base mounting which fits on top of instruments such as telescopic alidades. On many instruments, particularly theodolites, magnetic compasses can be built-in as an optional extra **6**. At least one manufacturer incorporates a magnetic device which automatically orients the horizontal circle. (See Chapter 8)
More accurate instruments are available such as the Wild double image prismatic compass **7** and the Wild polar attachment for theodolites etc, **8**. These instruments however provide a degree of accuracy not normally called for in site surveying.

4 *Francis Barker dry prismatic compass*

5 *Francis Barker military type of liquid prismatic compass*

6 *Wild orientation compass*

8 *Wild polar attachment for azimuth determination*

7 *Wild double-image prismatic compass*

Chapter 2·5

Surveying instruments: Measurement of height

This section describes instruments designed for the direct measurement of height or determination of levels

MEASUREMENT OF HEIGHT

Mechanic's level

Almost all instruments used for determining differences in height in surveying depend upon the use of spirit levels. The simplest is the mechanic's or builder's level **1** which can be used to measure small differences in height or level over short distances with the help of a long straight edge. This is normally adequate for checking over small areas but it is too laborious and inaccurate for larger scale work.

Reflecting levels

A pocket reflecting level **2** provides an instrument which can be used for rough levelling as, for example, in preliminary surveys. It consists of a tube through which an object can be sighted while at the same time the reflected image of a bubble shows the line of sight to be horizontal. Height of the observed level above ground at the observer's feet can be measured by holding the level against a staff or rod **3**.

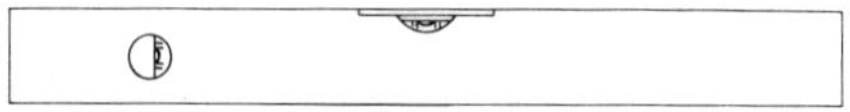

1 *Mechanic's or builder's level with two spirit levels*

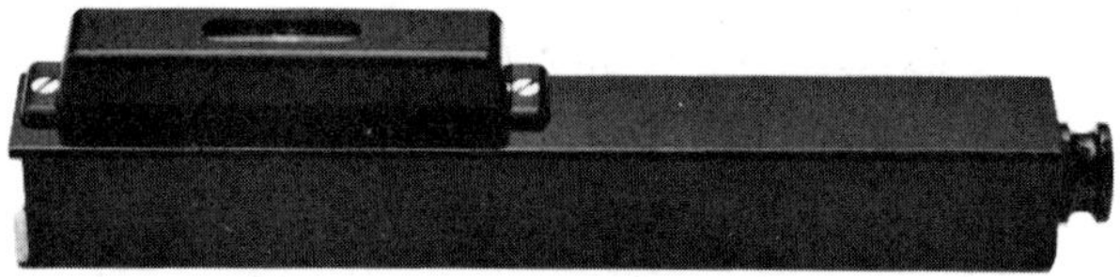

2 *Francis Barker pocket reflecting level*

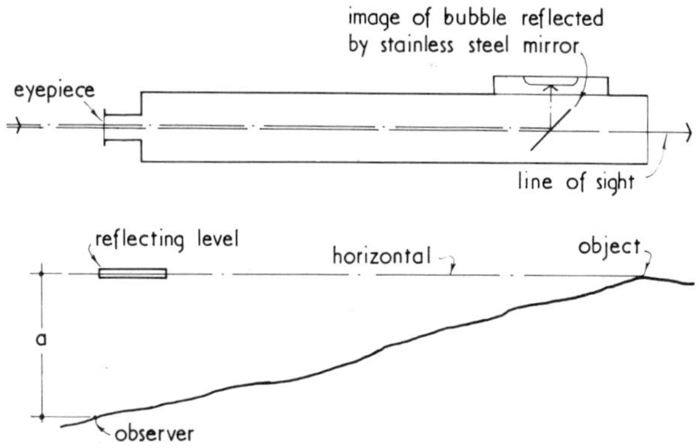

3 *Principle of reflecting level*

The Cowley level

For small scale surveying, such as drainage work and setting out buildings the Cowley level **7** is a very useful instrument for distances up to 60m and needs very little skill to operate it. The optical system employed is shown in **4**. It consists of two sets of mirrors, side by side so that the image seen in one set is displayed against the other, and when they coincide at the division line the line of sight is horizontal. The system depends upon a pendulum mirror which is suspended so that it remains in a horizontal plane whichever way the case is tilted (within reasonable limits). It is operated by moving a target bar **5** up and down against a graduated staff until both halves of its image coincide. The height of the horizontal bar can then be read off the vertical staff. The Cowley level is manufactured by Hilger and Watts and costs £28·50, complete with case stave and target.

Dumpy levels

The essential features of a dumpy level are illustrated in **8**, **9**. Most patterns have a telescope with a positive image (upside down) in order to give good definition and illumination, though modern instruments with bloomed lenses make it possible to have an erect image. Telescopes are mounted on

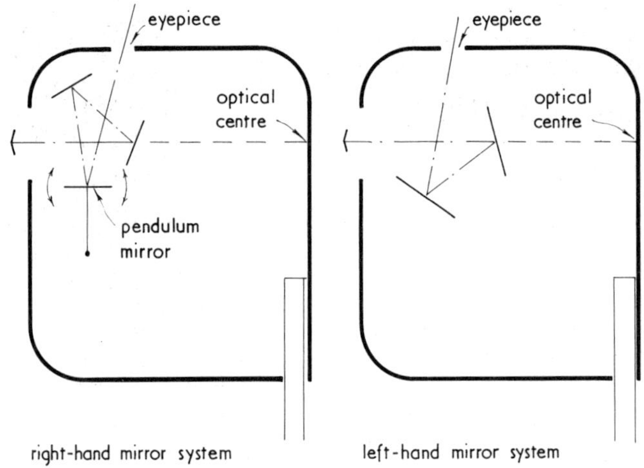

4 *Optical principle of Cowley level*

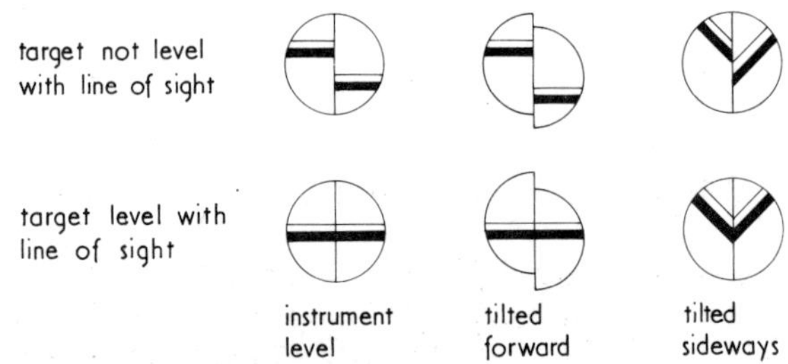

5 *Target seen through Cowley level*

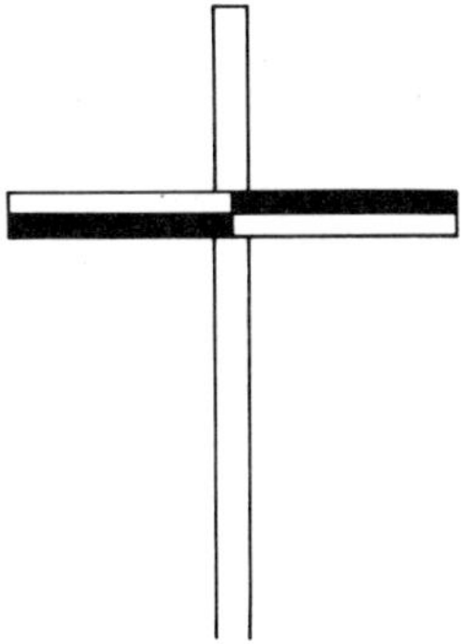

6 *Target used with Cowley level*

7 *Cowley level*

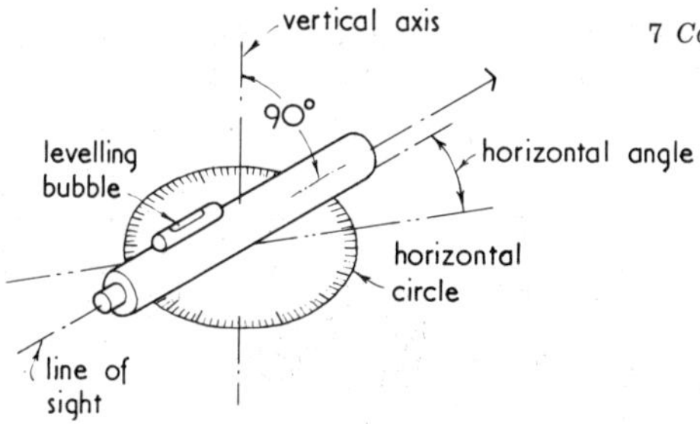

8 *Essential feature of dumpy level rotating in horizontal plane*

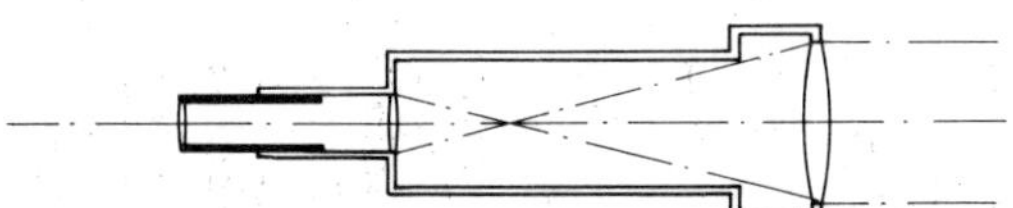

9 *Simplified section of level telescope to show projection of image on reticle*

and rotate about the vertical axis of a plate which can be set level by means of three (four in old instruments) levelling screws acting against a second plate (tribrach) which screws onto a tripod. When correctly levelled, by reference to a sensitive bubble, the line of collimation, ie the optical axis and line of sight through the telescope, moves in a horizontal plane. Proprietary examples are shown in **10** and **11**.

Levelling staves

By means of a graduated levelling staff **12** the distance downward from the line of collimation to the ground (or surface of some object) can be measured. The graduations on the commonly used Metric staffs are in metres, decimetres, centimetres and sometimes half-centimetres. In practice it is important that the staff should be vertical and so a small centre bubble is sometimes incorporated at the back.

Quickset levels

A modern development of the dumpy has a cup and ball mounting in place of the tribrach. This permits the telescope to be levelled approximately by means of a circular centre bubble. This type of level is known as a quickset.

10 *Wild Nk O1 level*

11 *Zeiss Ni 3 Builder's dumpy level*

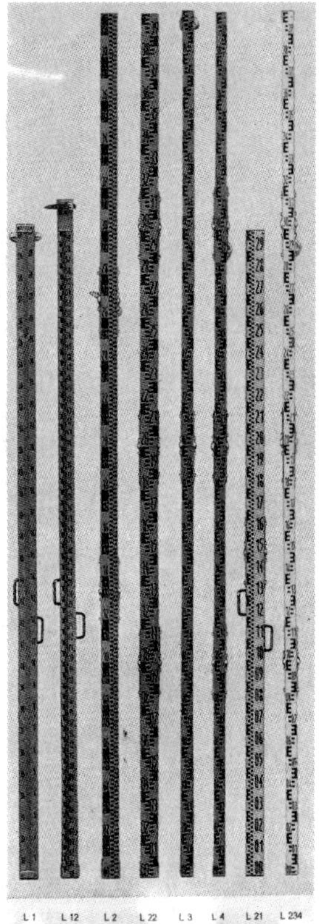

12 *Zeiss levelling rods, scale 1:20*

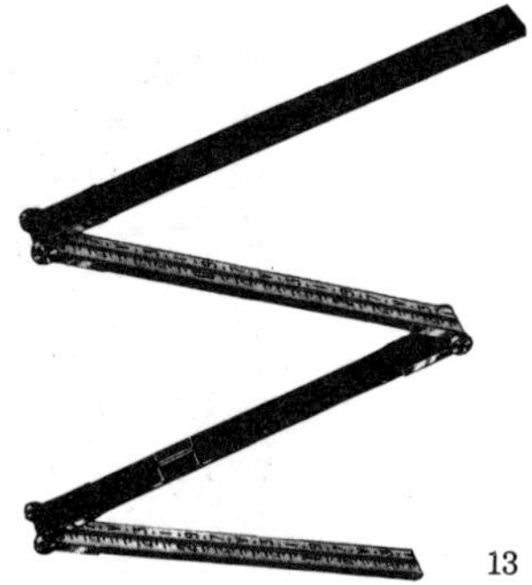

13 *Holbro four-fold levelling staff*

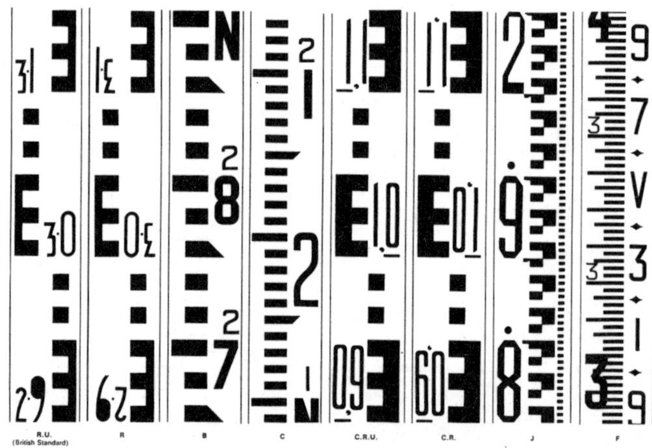

14 *Levelling staff readings—RU is British Standard metric reading*

Wye levels

These instruments differ from dumpy levels in that the telescope is fitted with two flanged collars which rest in two corresponding 'wyes' or hinged halved-rings which can be opened to allow the telescope to be rotated through 180deg about its optical axis, or taken out and replaced end to end **15**. The purpose of rotation through 180deg is that if the line of collimation is out of adjustment with the level bubble the error cancels out when the mean of two readings, one in each portion, is found **16**. Replacement end to end facilitates permanent adjustments to the instrument.

Reversible levels

In practice wye levels have not been popular because their construction is less rigid than that of dumpy levels and they are often defective in performance. However in modern versions of the wye level, called 'reversible', this difficulty has been overcome by encasing the telescope in a tubular mounting. Although this cannot be reversed end to end, the ability to rotate the collimation axis has been retained. A very high degree of precision in manufacture is necessary with these instruments because the level tube bubble must have the same centre when upside down, and the normal method of checking by reversal end to end is impossible.

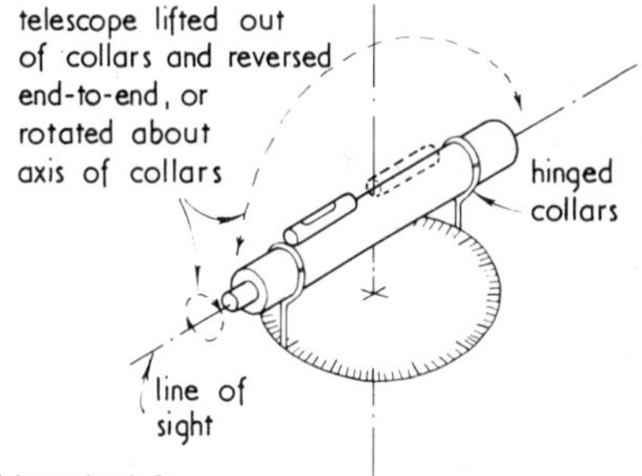

15 *Wye level in principle*

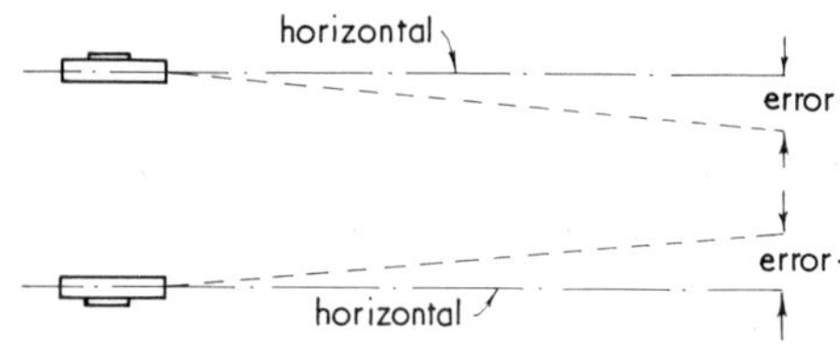

16 *Sighting with wye level in its normal and reversed projections*

Tilting levels

The essential features of a tilting level are illustrated in **17**. These levels differ from dumpy in that a screw is provided which makes it possible to adjust the line of collimation relative to the vertical

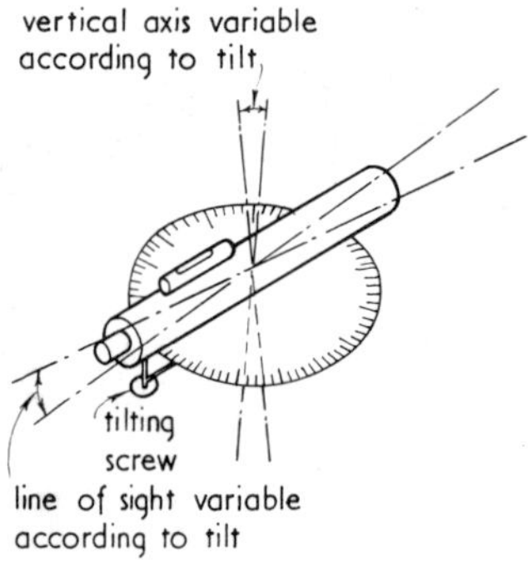

17 *Essential feature of a tilting level*

axis of the instrument. They may have a tribach or ball and cup mounting and are initially levelled approximately by means of a circular centre bubble. Fine adjustment by means of a sensitive bubble is made prior to each reading. The tilting screws are often graduated as gradienters. Some proprietary types are shown in **18**, **19** and **20**.

18 *Stanley A525 tilting level*

Stanley A450 quickset tilting level

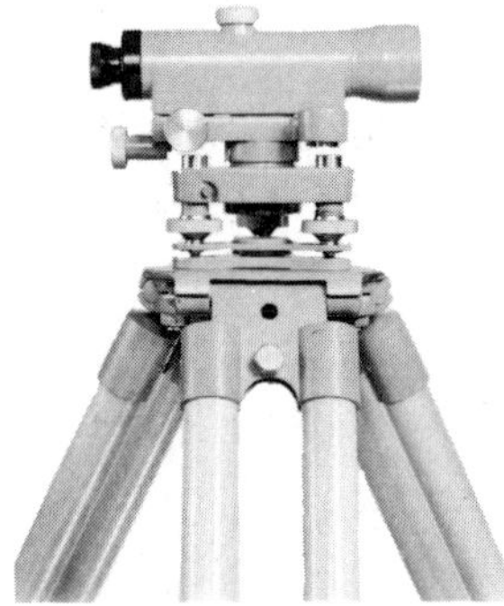

19 *Wild N 10 tilting level*

Hilger and Watts SL 10 tilting level

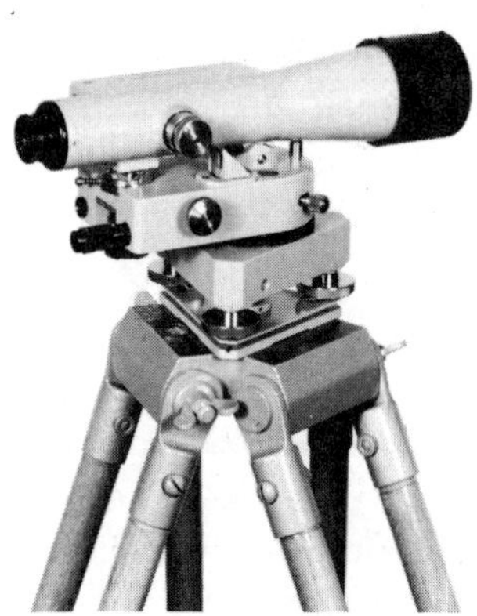

20 *Vickers s 440 tilting level*

Vickers s 23 tilting level

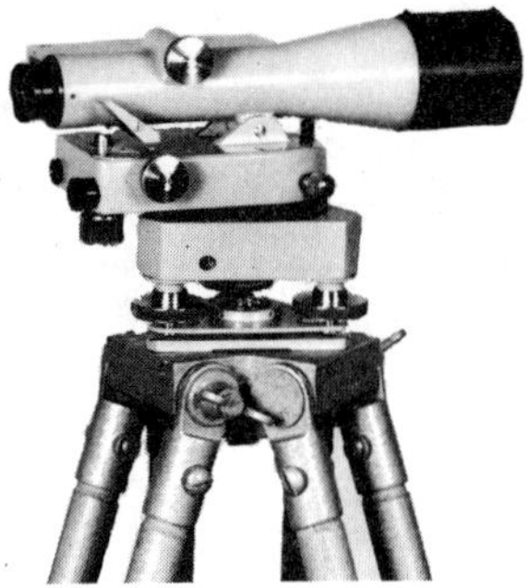

Vickers s 330 tilting level

Kern GKO tilting level

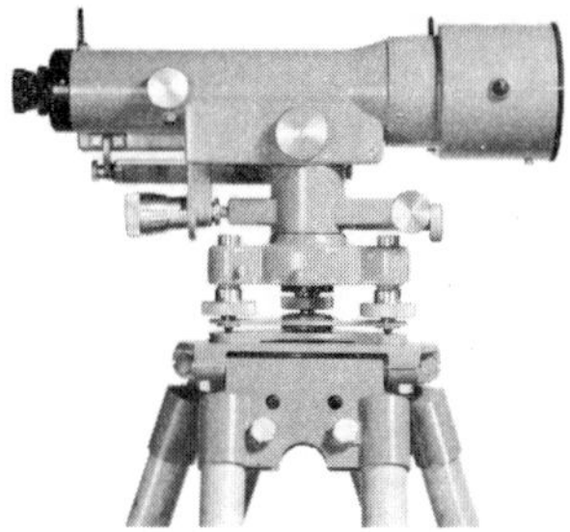

21 *Wild N3 precise level*

22 *Kern GK 23 precise level*

Precise levels

Precise levels are the aristocrats among levels and are used where the greatest accuracy is needed and for long distance work, eg structural deflection and first order geodetic surveying. They are usually distinguished by provision of optical coincidence level bubbles, large objective lenses, high magnification, a gradienter and an optical micrometer **21**, **22**.

Optical micrometers

Optical micrometers have a plane-parallel glass plate **23** set between the objective lens and diaphragm and controlled by a micrometer screw on the side of the telescope. This enables the operator to measure that displacement of the line of collimation necessary to bring it into coincidence with the nearest graduation on the staff. The micrometer must, of course, read zero when first sighting the staff. For work of this order of precision it is necessary to use a special invar staff, suitably graduated, since the optical micrometer reads to 0·05mm or 0·001ft.

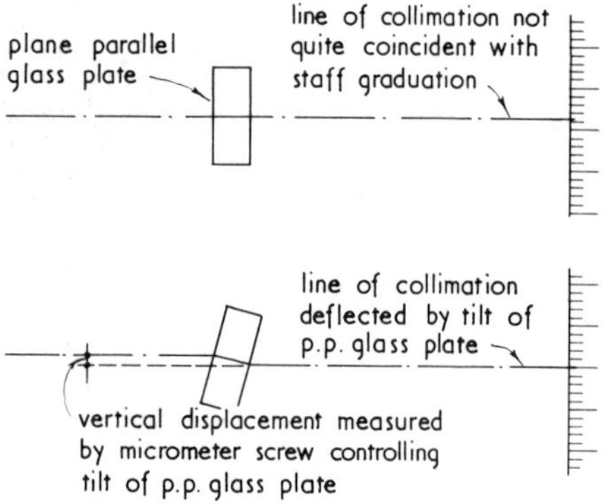

23 *Optical principle used for parallel plate micrometers*

24 *Stanley A 505 tilting reversible level*

25 *Wild N 2 tilting reversible level*

Tilting reversible levels

Tilting reversible levels **24**, **25** combine the main features of both tilting and modern counterparts of the wye level though they may have a tribrach or cup and ball mounting. They may be regarded as generally more precise than their parent types.

Automatic levels

An interesting recent development has been the introduction of automatic gravity-operated compensators in the optical system of otherwise conventional levels. The basic principle is illustrated in **26**. In practice prisms are used instead of mirrors, and several different methods of mounting them have been devised to avoid infringing comparable patents. Levels of this sort offer a high degree of accuracy with speedy operation. All of them require the instrument to be levelled within the limits of a circular centre bubble, but thereafter the line of collimation is corrected automatically. These instruments have an added value in that they are relatively unaffected by earth tremors or heavy traffic. Proprietary examples are shown in **27**.

The 'Aqualev' water level

When an existing building has to be surveyed it may

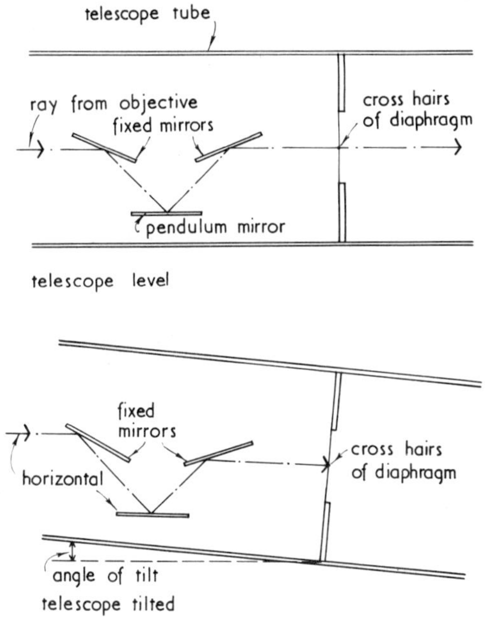

26 *General optical principles used in self-levelling levels*

27 *Automatic levels:*

Wild NAK 2

Kem GK1 A

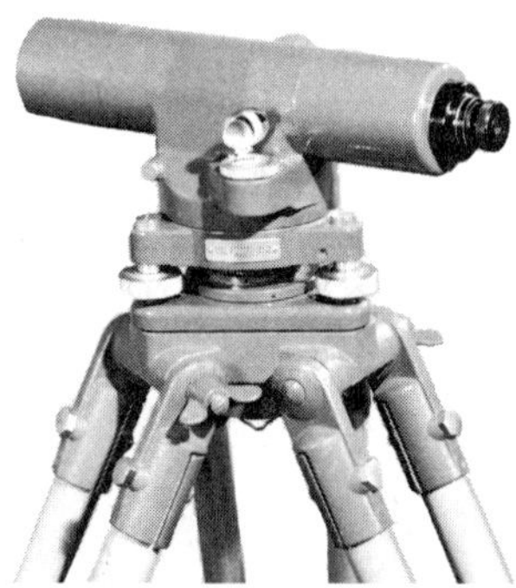

Hilger and Watts SL 60

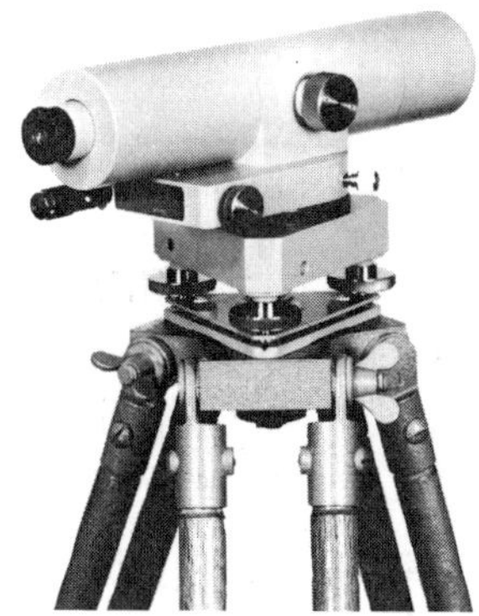

Vickers S77

Zeiss Ni 4 without graduated circle

Fuji AL-31

be profitable to use the 'Aqualev' water level. This simple but invaluable outfit enables levelling to be carried out in situations where other methods are unsuitable or impossible; eg, around, over or under obstructions. It is particularly useful for bricklaying, floor screeding, concrete shuttering, ceilings, steel erection, pipe runs, etc. The sight tubes are of unbreakable plastic with anti-airlock devices. For situations where it is inconvenient to sight surface of liquid against working surfaces, 'Aquarule' Scale Attachments are available.

28 *MDS Aquater level comprising 2 instruments of solid brass construction complete with 18·28m (60 ft) of flexible tubing*

Chapter 2·6

Surveying instruments: Measurement of horizontal angles

This section describes instruments or attachments to instruments for the measurement of angles in the horizontal plane. Chapters 3 and 6 deal with methods of reading, checking and handling such instruments

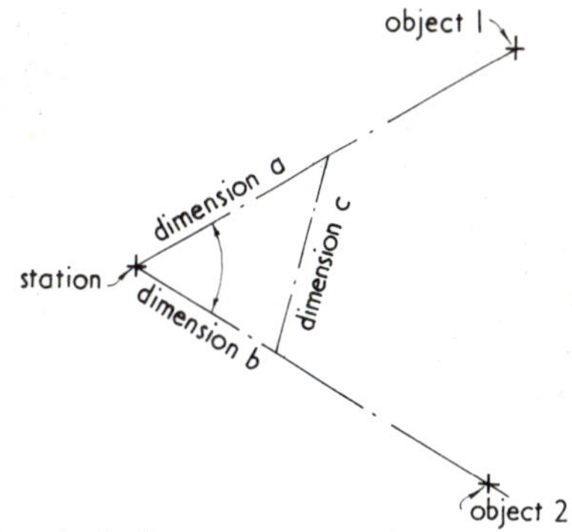

1 *Horizontal angles by linear measurement*

Chain or tape

The need to measure the horizontal or plan angle made at a point of observation between one object and another arises fairly frequently in surveying. This can be done by taped measurement as shown in **1** but not with any real accuracy. The method is as follows: measure a convenient distance (dimension a) along survey line to object 1; measure dimension b along survey line to object 2; measure distance between dimensions a and b (dimension c). This can be set up graphically in the drawing office and the angle measured. Alternatively it can be calculated. (See Chapter 9)

Magnetic bearings

Magnetic compasses can be used to determine angles by differences in compass bearings, but like taped angles this does not permit any very high degree of accuracy without considerable refinements of method. Compass bearings do, however, provide a very useful check when other methods of observing angles are used.

Plane tables

Plane tables are effectively drawing boards, measuring from 400mm × 400mm to 600mm × 750mm in size, supported on a tripod **2**. There is usually some provision for adjusting the plane table's surface into a horizontal plane; rotating it in that plane and clamping it at will against further rotation. The plane table is used in conjunction

with a sighting rule or alidade, a trough compass and a plumbing fork.
There are several methods of mounting plane tables on a tripod. The simplest is a tribrach which has a central socket on the board and three levelling screws as shown in **3**. More elaborate fittings incorporate cup and ball arrangements of which the Johnson movement **4** is perhaps the best when steadiness is important.

Alidades

An alidade is a rule with a sight wire attached used as a plane table for drawing a line in the direction of a sighted object. In its simplest form it consists of a wooden straight-edge with a vertical frame at each end having a wire stretched centrally from top to bottom so that it is possible to sight on distant objects **5**, **6**. In use a point is marked on the paper on the plane table and the alidade is placed against it and moved about until the required distant object is in the line of sight. A ray can then be drawn through the point and the procedure repeated for a second point as shown in **7**. An alidade with a parallel rule has been shown in the diagram. This is a normal provision on all but the most elementary alidades.
Simple alidades are sufficient for straightforward

2 *Holbro RE pattern plane table*

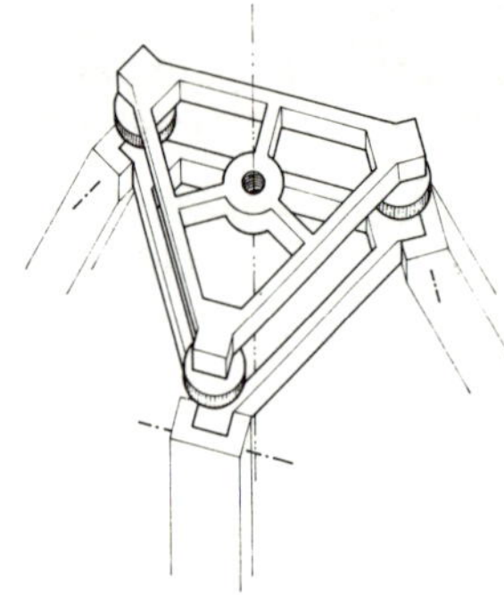

3 *Tribrack mounting for a plane table*

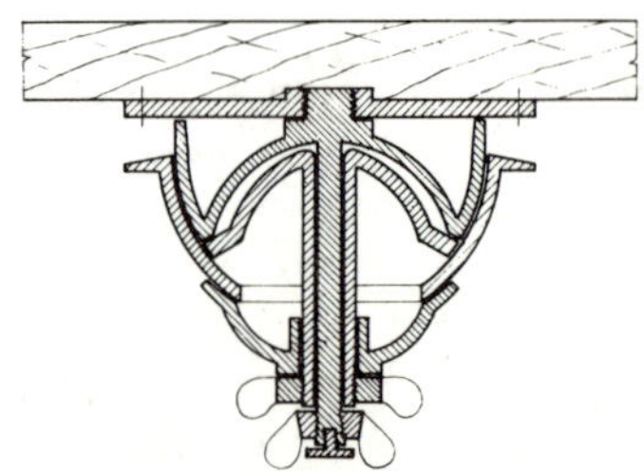

4 *The Johnson movement mounting for plane tables*

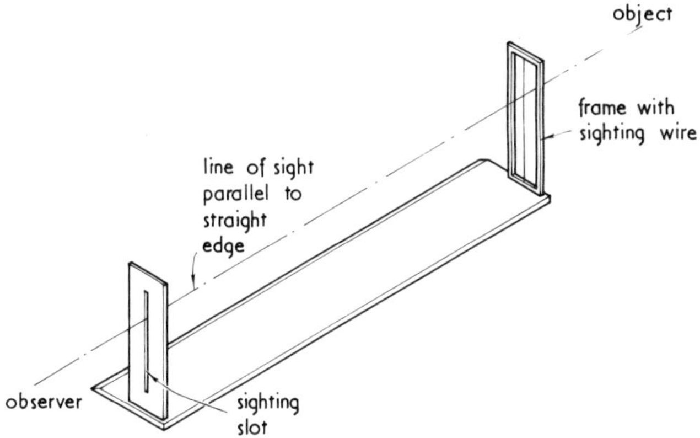

5 *Essential features of a plane table alidade*

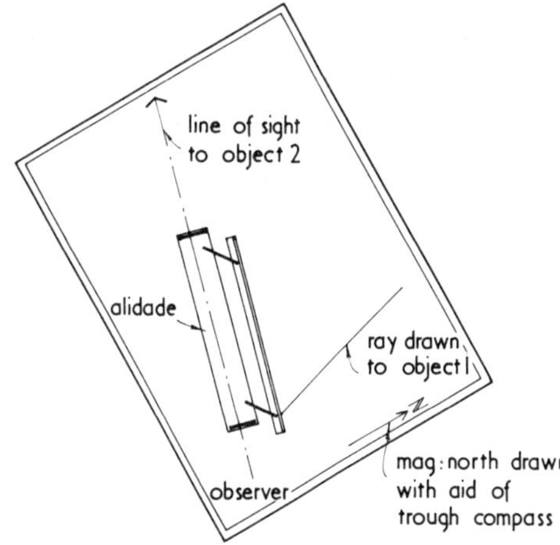

7 *Diagrammatic illustration of an alidade in use on a plane table*

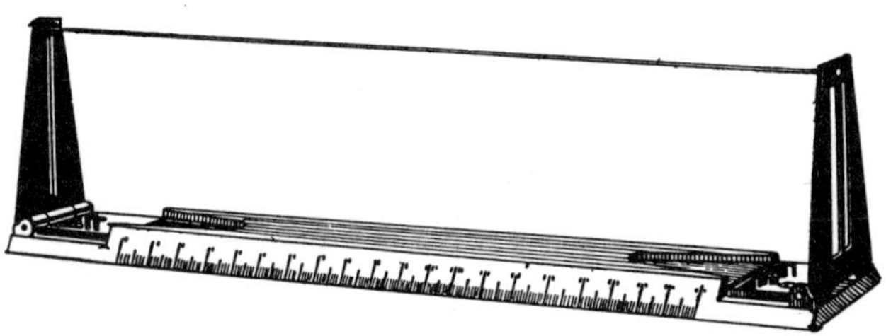

6 *Holbro metal alidade*

plane tabling, but when contouring is undertaken—for which a plane table is eminently suitable—work is much simplified and more accurate when a telescopic alidade is used.

Telescopic alidades

Telescopic alidades **8** are precision instruments which usually incorporate stadia lines, with a Beaman arc or some other reducing device for tacheometry and sensitive levels for levelling. The operator can plot each station point and its level straight onto the board with the help of an assistant who merely moves an ordinary levelling staff from point to point as required. Distance, direction and level are determined by the operator at the plane table through drawing and instrumental readings.

Horizontal circles

Many levels can now be obtained which incorporate a device for reading angles in the horizontal plane relative to the vertical axis of the instrument. The simplest kind usually takes the form of a chamfered edge to a metal circular base, graduated in degrees, which remains static and from which angles are read by means of a mark on an arm which moves with the telescope. More refined instruments have a vernier on the arm permitting readings down to about 5min to be made **9**.

The advantages of being able to read horizontal angles with a comparatively simple instrument such as a level have been sufficient to cause more and more manufacturers to incorporate more elaborate, accurate and finely graduated horizontal circles. On some instruments these are now engraved or etched on glass, totally enclosed and read by optical means through an eye piece.

8 *Wild RK1 Telescope alidade*

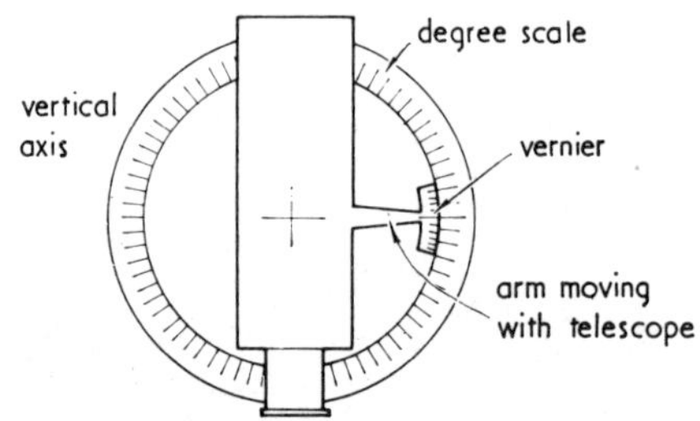

9 *Level with a horizontal circle and vernier index*

Chapter 2·7

Surveying instruments: Measurement of vertical angles

This section describes instruments for measuring vertical angles, and is therefore largely concerned with theodolites. The subject is expanded in Chapters 3 *and* 6

Abney levels

One method of finding the difference in height between two or more points is to measure the vertical angle (ie the angle from the horizontal) between two points and calculate. One of the simplest instruments for the purpose is an elaboration of the reflecting level, called the Abney level **1**. It employs a similar tube but with the spirit level mounted on the cross-axis of a graduated semicircle and fixed to a perpendicular index arm so that it becomes possible to measure vertical angles from the horizontal and hence heights or gradients. The arc is usually graduated with a gradient scale as well as giving angles down to 10min. Refinements such as verniers and draw tube telescopes are normally incorporated **2**.

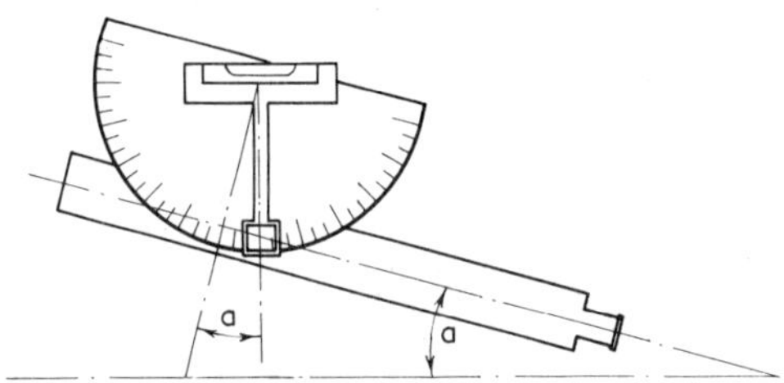

1 *Essential features of an Abney level*

Astrolabes and sextants

Astrolabes and sextants **3** are instruments for measuring vertical angles, but since their use is chiefly for astronomical observation they are not covered here.

Theodolites

The best known and most widely used instruments for measuring both vertical and horizontal angles are theodolites **6**, **7**. These instruments can serve also for ordinary levelling and generally have stadia lines incorporated for tacheometry. Magnetic compass needles are either built-in or supplied as an accessory. The general principles of theodolites are given in **4** and **5**. The cost range of theodolites is considerable because at one extreme they are comparatively simple and of sufficient accuracy for comparatively small surveys and

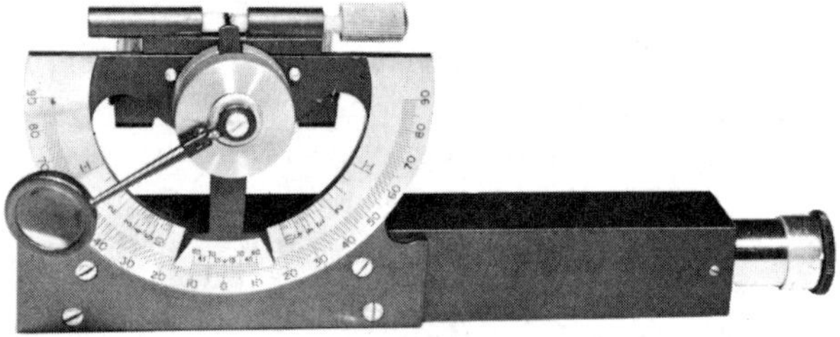

2 *Francis Barker Abney level*

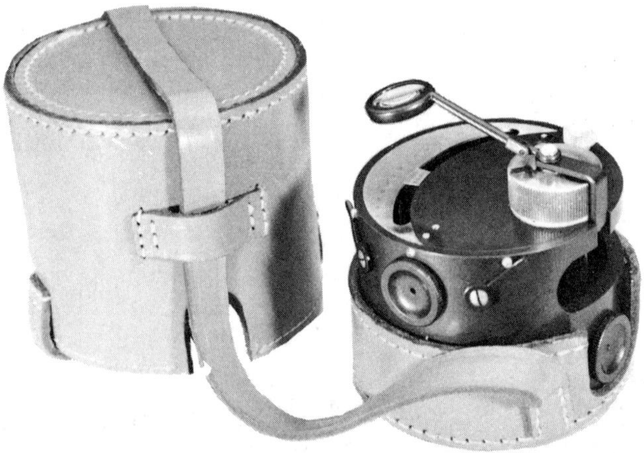

3 *Francis Barker box sextant*

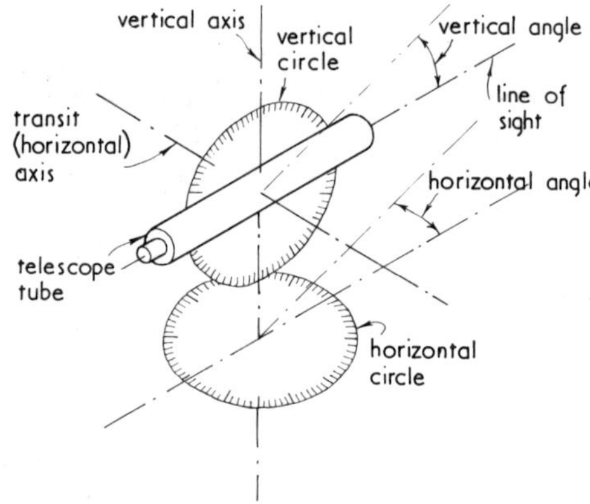

4 *Essential features of theodolite*

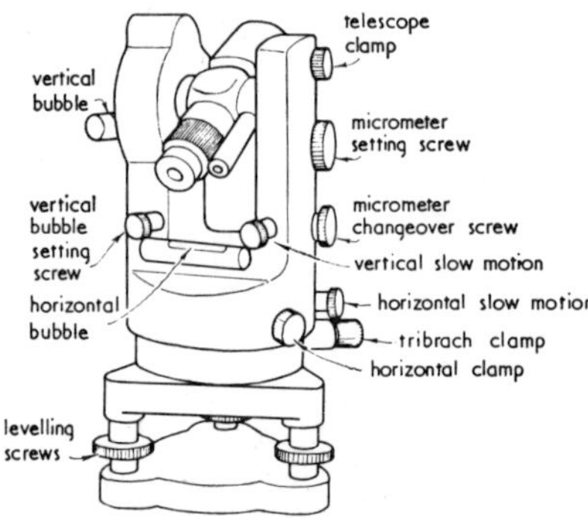

5 *Main adjustment controls of theodolite*

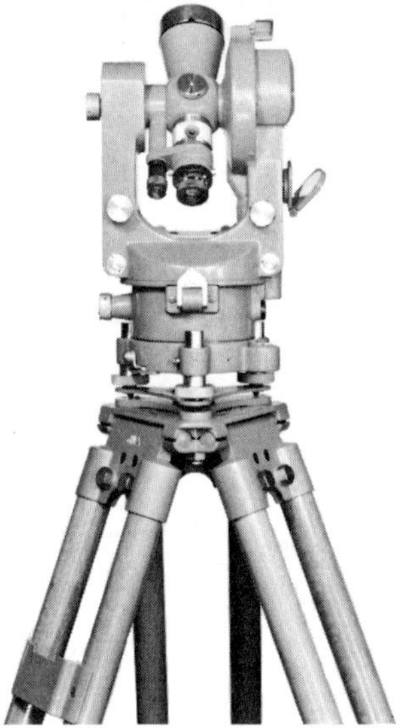

6 *Wild T 16 theodolite*

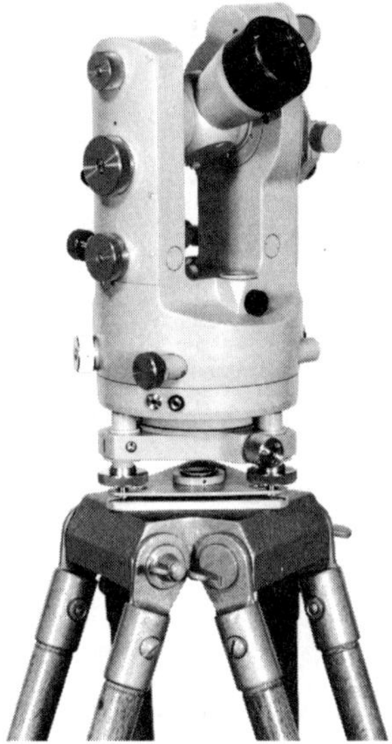

7 *Vickers v 406 theodolite*

setting out, while at the other they represent the most precise instruments available for large scale survey embracing countries and continents. This chapter is confined to types likely to be used in general practice. All theodolites have certain points in common although the same ends are met by a variety of means. A number of proprietary types are illustrated at the end of this chapter.

Theodolite telescopes

Salient points are as follows:

Older instruments were fitted with externally focusing telescopes but their many disadvantages, despite their simplicity, have caused makers to adopt the internally focusing telescope without exception. A common arrangement is shown in **8**.

The *diaphragm* carries the *reticle* (the words are sometimes used synonymously) which in modern instruments is a glass plate engraved with extremely fine lines in a pattern permitting horizontal and vertical intersection of an object seen through the telescope. See **9**.

The telescope is mounted on a horizontal axis—the transit axis—the ends of which are held in a fork called the trunnion **10**. The telescope can rotate through a full circle, the operation of turning it through 180deg being termed 'transiting'. (See Chapter 6)

Eye pieces are fitted with one or two dark glasses which can be pulled over the lens for viewing the sun. *On no account should the sun be sighted without*

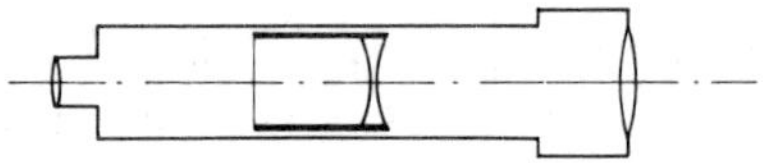

8 *Internal focusing arrangement in a theodolite*

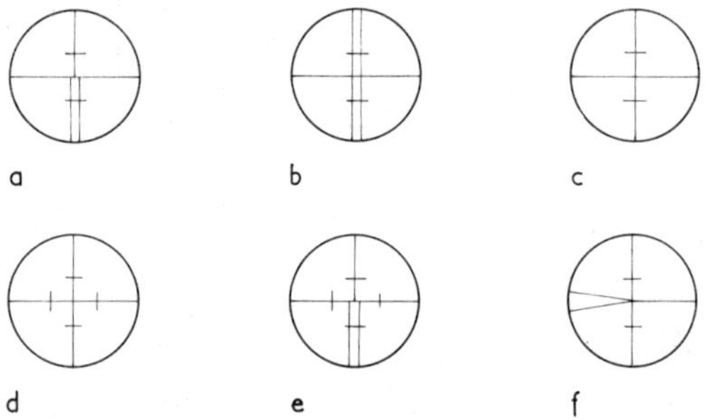

9 *Common patterns of reticles in theodolite diaphram*

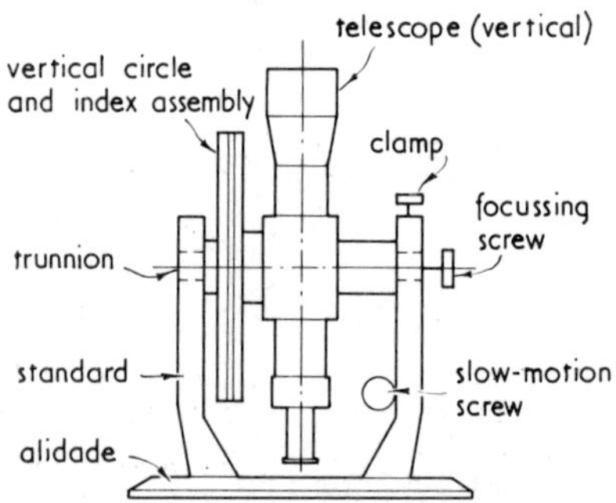

10 *Main features of theodolite telescope mounting*

a dark glass in position because the concentration of the sun's rays will destroy the eye's vision permanently.

The vertical circle One end of the transit axis is usually connected to a graduated circle which moves with the telescope **11**.
The transit axis extends through an index assembly which does not rotate but has a sensitive spirit level mounted on it exactly parallel (if the instrument is correctly adjusted) to the index marks **12** from which vertical angles can be read.
A screw working against a spring plunger enables the spirit level to be levelled before readings are taken so that the index marks are truly horizontal. Fine adjustment to bring the telescope exactly on to the object (target) is not easy when it is moved up and down by hand, and so a clamping screw is provided which fixes the telescope so that final adjustment can be made with a slow-motion screw. These are usually in the position shown in **5** on the same side as the telescope focusing knob.

Theodolite alidades

The telescope, vertical arc and index assembly are supported on trunnion standards and the standards themselves are mounted on the alidade **13**, which in

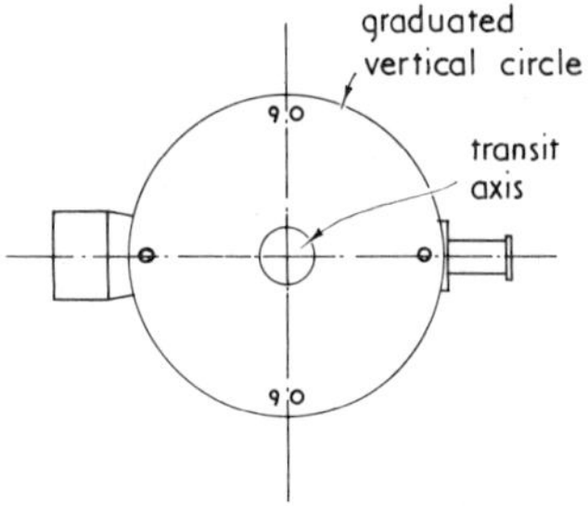

11 *Vertical circle*

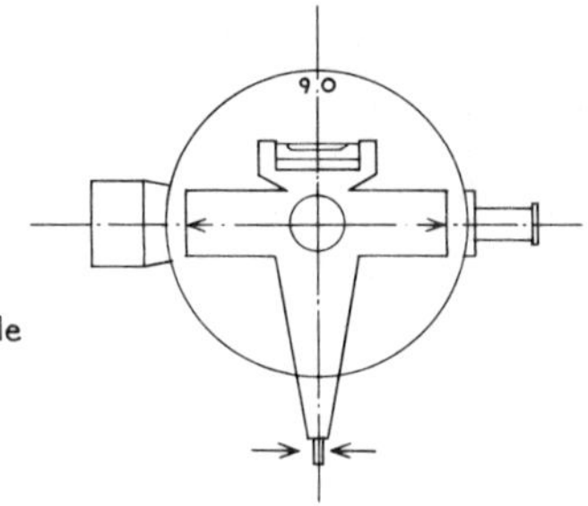

12 *Index assembly*

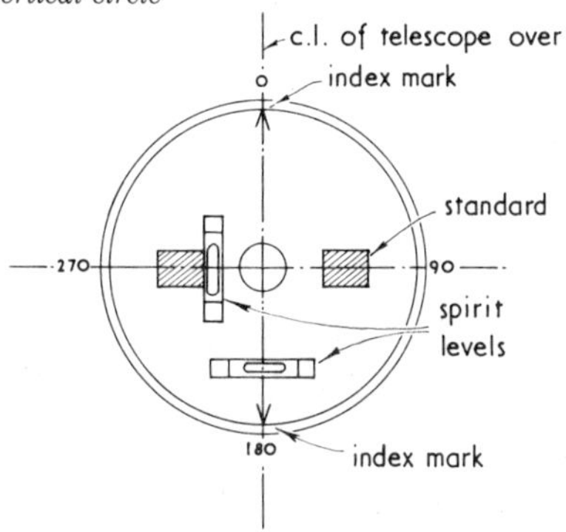

13 *Theodolite alidade in sectional plan view*

this case is a circular metal plate which has on it the reading indices for the horizontal circle and one or (more usually) two spirit levels. The second is less sensitive than the first and is there to simplify the task of levelling the alidade. The alidade is free to rotate horizontally on a pivot, usually conical, which centres on the vertical axis of the theodolite.

Theodolite horizontal circles

The horizontal circle is a second plate, also free to rotate, which fits round the alidade as shown in section **14**. A clamping screw clamps the alidade and circle together when required, and there is also a slow-motion screw. The latter controls the rotation of the one relative to the other: thus the index mark can be set against any graduation (through 360deg) in the horizontal circle.

The tribrach

Both the horizontal circle and alidade are together free to rotate relative to the tribrach **15** which contains the pivot socket. The tribrach is usually an equilateral triangular plate with the pivot socket at its centre. It is fitted with a clamping screw which clamps the horizontal circle at any point in its rotation that may be desired and brings into action a second slow-motion screw for fine

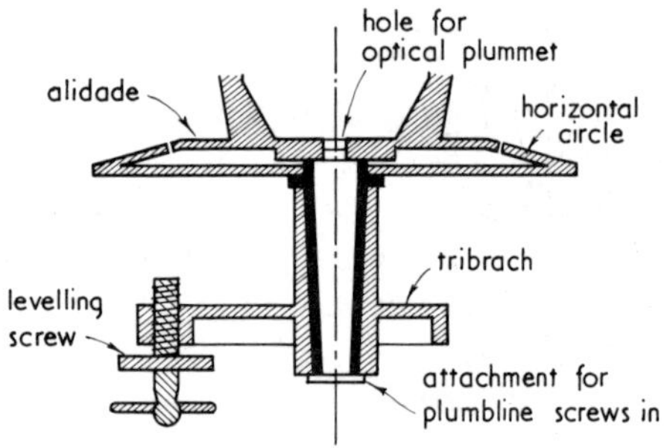

14 *Theodolite horizontal circle and its mounting*

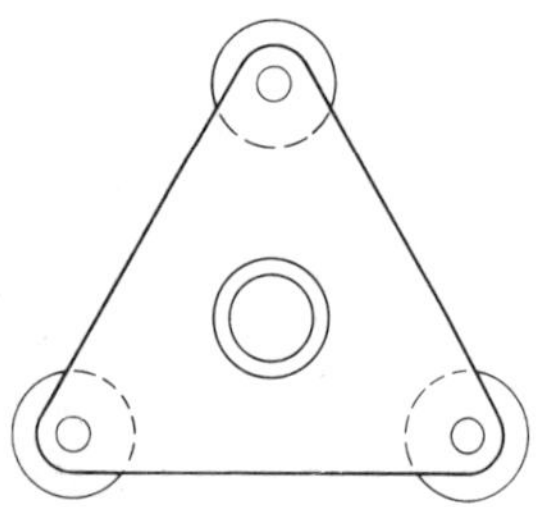

15 *Theodolite tribrach*

adjustment. This latter screw usually has its head milled in some way that distinguishes it from the alidade clamping screw. On modern types of levels the tribrach is supported by three levelling screws which are, in turn, connected to the *trivet stage*.

Trivet stages

The construction of trivet stages varies a good deal but the general principle is illustrated in **16** where it can be seen that the bottoms of the levelling screws are held by a ball and socket arrangement.

Adjustable heads

The trivet stage is normally mounted on a plate or *movable head* over which it can slide within the limits of a circle of about 50mm diam. This enables the theodolite to be centred over the ground point (nadir) before the trivet stage is tightened down.

Tripods

The movable head screws down on to the tripod. The latter closely resembles the kind used for levels but is usually more robust and may have a circular bubble incorporated to facilitate levelling before the theodolite is screwed on. The legs of modern instruments are telescopic.

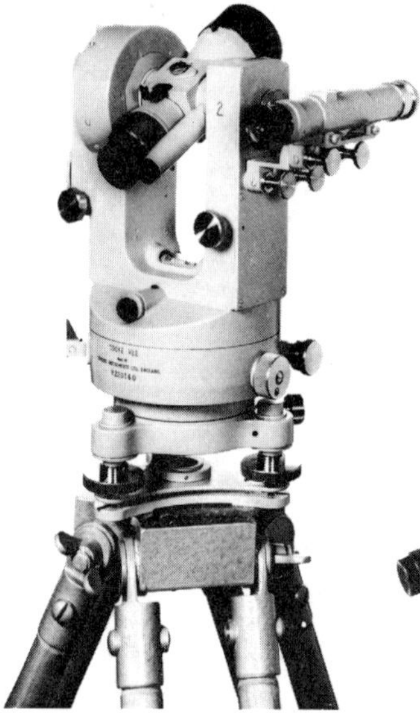

Vickers V22 scale reading theodolite

Stanley A1 theodolite with optical micrometer

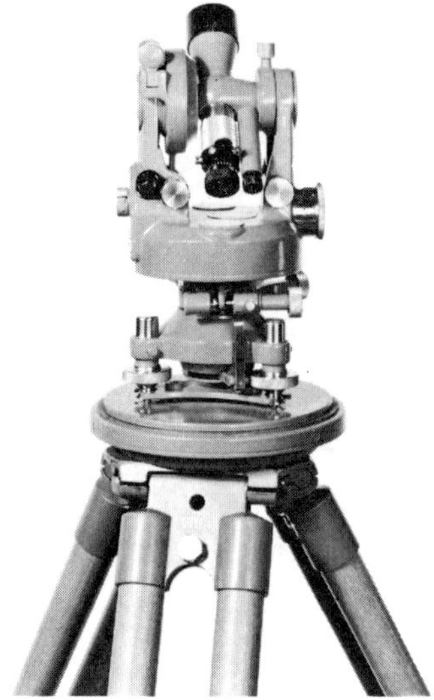

Wild T0 builder's theodolite

Wild T2 theodolite for high accuracy work

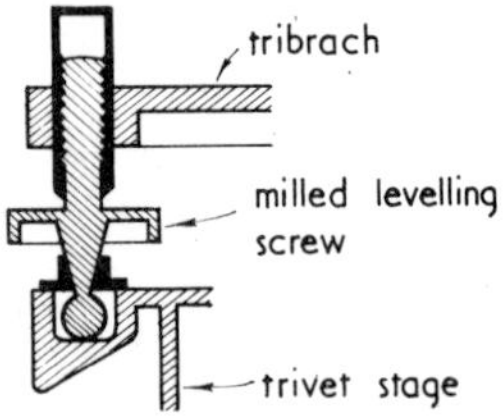

16 *Mounting of theodolite tribrach on trivet stage*

Kern DKM 2 theodolite with optical micrometer

Kern DKM 1 engineer's theodolite with inverting telescope

Wild T1A double centre theodolite with optical plummet

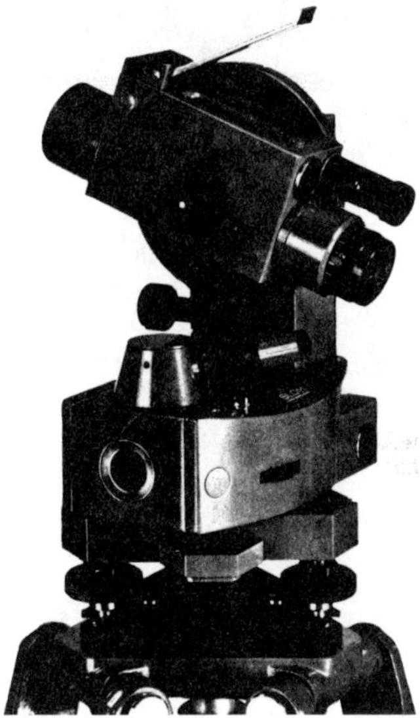

Slom STNO theodolite

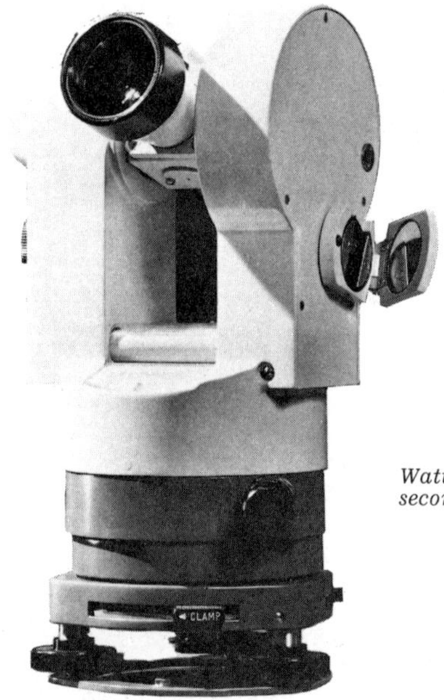

Watts ST 310 microptic 10 second auto indexing theodolite

Watts ST 200 1 second theodolite

Chapter 2·8
Surveying instruments: Measurement from photographs

This section briefly describes photogrametry and some of the instruments used in collecting data from photographs

General
Mapping from photographs started in about 1850, almost as soon as photography itself. The techniques used in photogrametry can be traced back even further to the time of the camera obscura.

Terrestrial Photography
Measurements from terrestrial photographs can assist in construction projects of all kinds. Surveys for measuring the volumes of excavations or of spoil heaps etc, studies of landslides, avalanches, as well as uses in architecture, archeology, and the conservation of cultural relics. The terrestrial camera may be used alone or with a theodolite **1** and **2**.

Aerial Photography
The first aerial photographs were taken from station balloons using conventional hand held

1 *Wild p32 Terrestial camera*

2 *Wild p30 photo theodolite*

cameras. The developments in technique and apparatus of aerial photographs the preserve of a few specialist companies, notably the Ordnance Survey and Hunting Surveys **3** and **4**.

Stereo Pairs

Aerial Photographs are taken from an aeroplane flying a carefully pre-determined flight pattern in which the flight path, direction, height and speed are carefully controlled. This ensures that strips of photographs are taken covering the whole area under survey. Adjacent photographs in each strip overlap one other by 60% ie 60% of the ground in one photograph also appears—from a slightly different point of observation—in the next photograph. Such a pair of photographs form a stereo pair, and when viewed through a stereoscope give an impression of depth that is utilised in producing contour maps.

Photographs are rectified ie corrected for faults in the flight control, and enlarged to a suitable size **5**.

Stereoscopes

The simplest stereoscopes are the folding mirror type **6**, that only cost about £2 and are only suitable for qualitative work. A more elaborate stereoscope may be used with a parallex bar to make spot height

3 *Wild RC 10 Universal film camera installed in a Scottish Aviation 'Twin Pioneer'*

4 *Photographic flight with an automatic aerial camera The overlap of successive vertical photographs is 60%*

measurements and to prepare photographs for plotting **7**.

Plotters

Plotting maps from stereo pairs is done most efficiently by plotters **8**. A plotter may be able to scan a pair of photographs and automatically produce a high definition contour map of the overlap area or store the information in digital form suitable for storage or computer analysis and manipulation.

Maps

An example of the main stages in the production of a map from aerial photographs is given in **9**. This shows firstly part of an unrectified 180mm × 180mm photograph taken from a height of 2·5km above ground. The rectified photograph has been enlarged from a scale factor of 1: 22 000 to 1: 5000. A photogrammetric plot of the enlarged photograph gives a good indication of the standard of mapping possible.

Costs

It has been estimated that for large scale work a photogrammetric survey will be 40% cheaper than a conventional geodetic survey carried out on the ground with theodolites levels chains and tapes. This takes no account of the saving in time which is considerable.

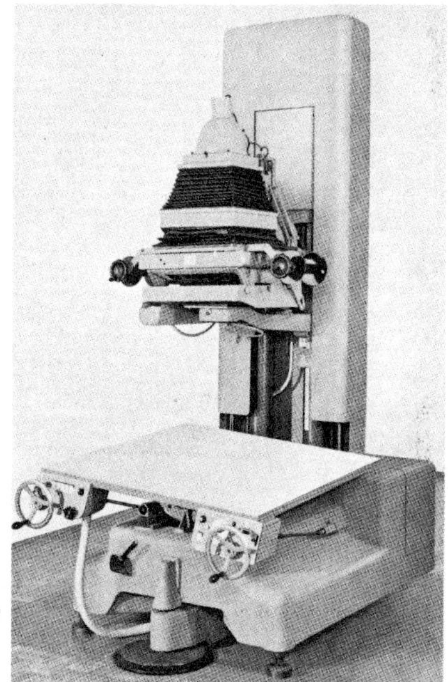

5 *Wild E4 Rectifier-enlarger*

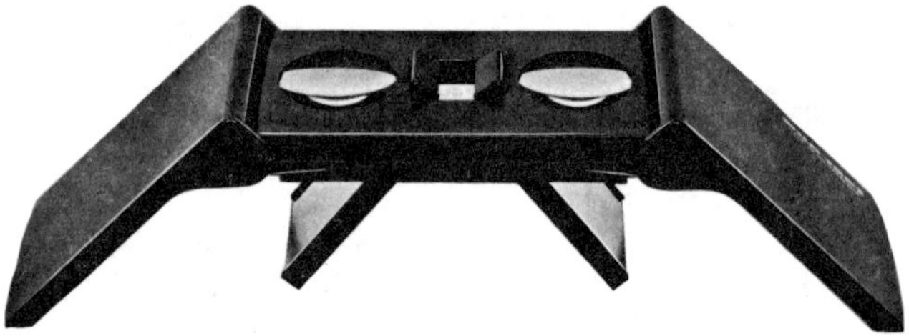

6 *Wild pocket mirror stereoscope*

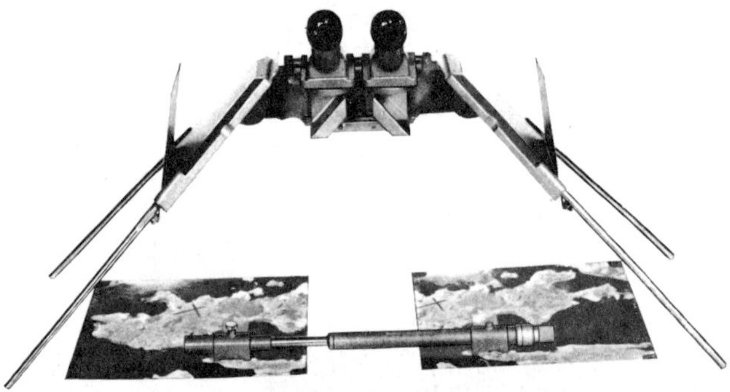

7 *Wild ST4 mirror stereoscope with binocular attachment and parallax bar*

8 *Thompson Watts Plotter model 2*

1

2

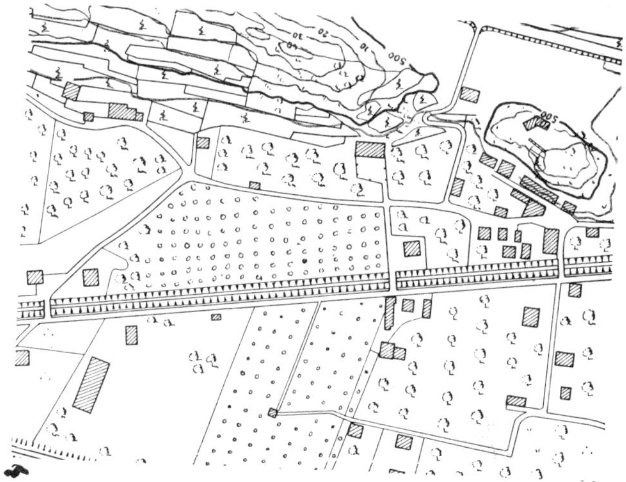

3

9 *Example of plotting*

1 Part of an original photograph reproduced at 1:1 Wild Ariogon lens f = 11·5cm. Format 18cm × 18cm. Flying height above ground 8200ft. Picture scale about 1:22000
2 Small picture enlarged to the scale 1:5000
3 Photogrammetric plotting at 1:5000

Chapter 3
Surveying instruments: Accuracy and angular scales

This chapter gives guidance on standards of accuracy to be expected from surveying instruments in general use, so that an appropriate choice can be made for a particular job. The chapter also considers the various types of angular scales used for surveying instruments and the way in which they are read

General

In surveying practice the instruments used for any particular project should be chosen in accordance with the degrees of accuracy that are appropriate. These will vary according to circumstances, but the chief factors that operate are land value and size. The price of urban land is such that a difference of 5mm in length of frontage makes an appreciable difference in cost, whereas in the country a tolerance of 10cm might be accepted.

So far as size is concerned errors tend to magnify proportionately, hence instruments of greater precision become necessary as distances to be measured increase.

This chapter gives some idea of the order of accuracy to be expected from instruments in each functional group. The data given should, however, be compared with the technical data given by manufacturers for specific instruments. Some of these are tabulated in more detail in Chapter 8.

Table I *Accuracy of instruments: Vertical orientation*

Instrument	**Accuracy to be expected**	**Comments**
Plumb line	± 3mm	Suspended about 1m above ground in good weather with no wind
Circular level	8min per 2mm for good instruments	Varies considerably
Tubular level	10 sec per 2mm up to 30sec per 2mm for precision levels	
Optical	1mm in 10m to 1mm in 30m	Higher accuracy can be obtained with autoplumbing devices

Table II *Accuracy of instruments: Linear measurement*

Chain	Error of about 1 in 500 down to 1 in 100	1 in 500 represents a chain in fairly good condition, 1 in 1000 is the utmost accuracy attainable. These figures assume that no chaining mistakes are made

Instrument	Accuracy to be expected	Comments
Steel bands and steel tapes	Errors between 1/21 000 for short lines to 1/25 000 for long lines may be expected	These figures assume that every normal refinement in taping is practised and that tapes are standard within 1 in 100 000 For detailed exposition of sources of error see CLARK, D. Plane and geodetic surveying for engineers, vol 1. London 1961, Constable, 5th edition revised by J. Glendinning
Linen tapes, metallic and non-metallic	Variable	Should not be used where precise results are needed unless constantly checked against a steel tape. They stretch when pulled and shrink when wet
Rangefinder	Errors approximate to 200mm at 200m and 1m at 500m	These instruments are useful chiefly for airfield work or where long distances are involved
Tacheometers or telemeters and levels or theodolites with stadia lines	Error in traversing should not exceed 0·06 $\sqrt{\text{Distance in m}}$ both for linear and height closing error. Accuracy is usually at its best for sights between 120m and 200m. A modern double image reduction tacheometer can improve performance by about ten times the above figure ie to 0·006 $\sqrt{\text{Distance in m}}$	The figures given represent an average of combined sources of error: reading staff intercepts; instrument errors; natural causes such as wind and atmospheric refraction
Invar subtense bar and horizontal arc	Accuracy depends upon the quality of instrument used and observational technique employed. An accuracy of 1/10 000 is not unusual for high grade work	More accurate than is usually necessary for architectural work

Table III *Accuracy of instruments: Right angular measurement*

Cross-staff heads	Within 0·1m at 5m	Not very accurate but quite adequate for long offsets
Optical squares	Within 0·02m at 5m	Much depends upon the care with which they are handled

Table IV *Accuracy of instruments: Orientation*

Magnetic compasses	The accuracy of magnetic compasses may be assessed by the scale provided to read them	Local factors can make a great deal of difference, such as presence of iron, annual and diurnal variation etc. High precision is usually beyond the needs of architectural practice

Table V *Accuracy of instruments: Measurement of height*

Instrument	**Accuracy to be expected**	**Comments**
Mechanics level	5mm at 25m	The accuracy of results obtained depends very largely on the skill and care with which they are used. The figure given represents considerable care and expertise
Reflecting level	Convenient rather than accurate	Intended for approximate work and rough checking
Cowley level	About 5mm at 25m	Accuracy increases for shorter distances, decreases for longer
Dumpy level: *small*	Accuracy of single levelling (18×) 5mm to 6mm per km or 0·05ft per mile	These figures represent the greatest accuracy obtainable with good modern instruments
medium or tilting	(30×) ±5mm per km or 0·025ft per mile	In practice other factors come into play and the following
reversible	(24×) ±2·5mm per km or 0·012ft per mile	table represents reasonable performance:
automatic	(30×) 2mm per km or 0·010ft per mile	Error per 1km 10mm Error per 4km 20mm Error per 9km 30mm
precise	(42×) ±0·25mm per km or 0·001ft per mile The magnification (in brackets) is given as an indication of size of level Accuracy can be nearly doubled by double-levelling	Note that the distances given are those between start and extreme point, not the total distance traversed. The formula is: permissible error in mm = 10 $\sqrt{}$ distance in km. In favourable conditions it would be reasonable to expect the factor 10 to be reduced to 5

Table VI *Accuracy of instruments: Measurement of horizontal angles*

Plane table	Ordinary plane tabling is not very accurate, especially in this country where conditions are often unfavourable. For recording horizontal angles at a station point it is rather more accurate than chain measurement. When using telescopic alidades with a magnification of about 25×, accuracy can be expected to conform to the accuracy limits for tacheometry and levelling already noted	The kind of plane table surveying likely to be done in connection with architectural practice would be to scales of 1/2500 to 1/100. At these scales plane tabling should be limited to filling in detail to maps or plans prepared by other means. It is a useful quick and reliable technique for contouring or plotting curved or irregular shapes if a telescopic alidade is used
Horizontal circles fitted to levels	Accuracy in terms of nearest reliable estimation: (18×): to the nearest 30min (20×) to (42×): to the nearest 5min	The accuracy of horizontal circles fitted to modern levels is sufficient to obviate the need for a theodolite in most architectural practices unless a high degree of accuracy is required (as in London)
Horizontal circles fitted to theodolites	(20×): to the nearest 1min (27×): to the nearest 10sec	The figures given include only those instruments likely to be encountered in normal practice

Table VII *Accuracy of instruments: Measurement of vertical angles*

Instrument	Accuracy to be expected	Comments
Abney level	Up to 10min of arc is possible	Most accurate within the range of angles from 45deg depression to 45deg elevation. Much depends upon the operator, but the potential accuracy is unlikely to be achieved in full
Theodolite vertical circles	(20 ×) : to the nearest 10min (27 ×) : to the nearest 10sec	See previous note on theodolites. Some manufacturers will supply theodolites without vertical circles. These are cheaper and meet the need to measure horizontal angles above or below the line of collimation

Horizontal and vertical angles in practice

The foregoing tables indicate the accuracy that may be expected in reading instruments. This is not the same thing as the accuracy to be expected when using them in the field. The computation of probable angular error is something that has to be done for individual instruments. It is too complex to be noted here and unlikely to be of importance in ordinary architectural surveying.

ANGULAR SCALES

Reading optical instrument scales

Levels fitted with horizontal circles, theodolites, telescopic alidades, Abney levels and tacheometers have a variety of means provided for reading angular scales. These may be classified as follows:

1 direct and by estimation;

2 direct with vernier;

3 indirect with vernier;

4 indirect and by micrometer;

Direct scales are those which are wholly exposed to view: indirect scales are totally enclosed and viewed by optical means through an eyepiece.

Simple direct scales almost invariably have two index marks at 180deg apart. They are usually subdivided into 360deg main and either 30min or 20min inferior graduations. They can thus be read with certainty to the nearest 30min or 20min or by estimation to 15min or 10min **1**.

More accurate instruments with direct scales are divided and placed as in **1** above, but are provided with vernier indices and, usually, microscope eyepieces for reading them. Verniers all work on the same principle, but several varieties are to be found. It is fairly easy to make mistakes in reading verniers, especially when they are novel. Common

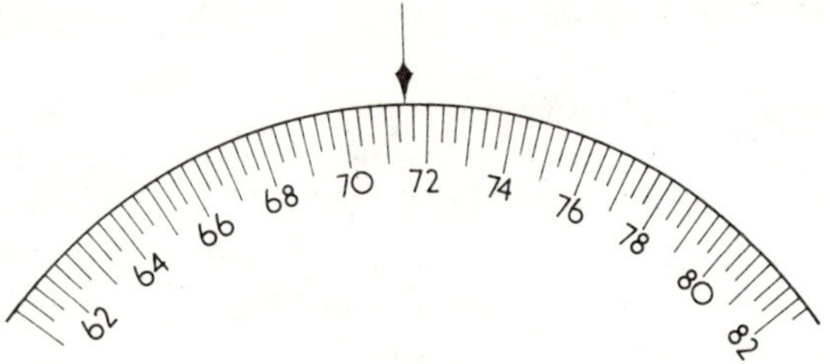

1 *Direct angular scale*

varieties which may be encountered on surveying instruments are given in **2** and **3**.

In practice it is sufficient to note the point where the vernier index (zero) falls relative to the preceding whole number of the main scale—in the first example, 10—and then count along the vernier divisions to find where one of them coincides with a division on the main scale—in the second example, 7. Thus the reading is 10 plus 0·7.

Retrograde verniers

Some verniers are made with their scale running in the direction opposite to that of the main scale. In such cases the vernier scale is made to cover eleven units of the main scale so that each of its units = $\frac{11}{10}$ = 1·1 of the main scale. In the example **4** co-

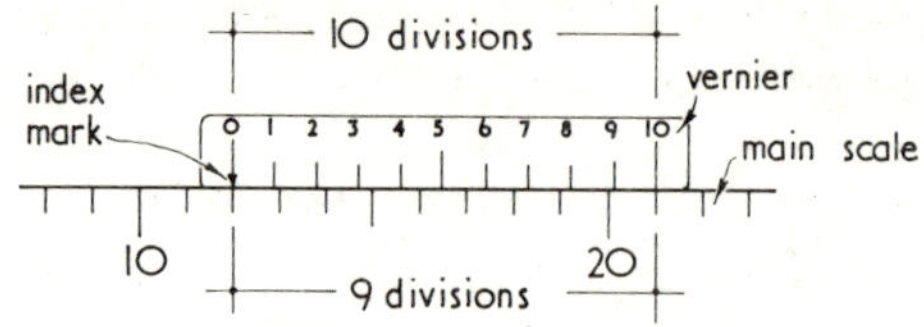

2 *Direct vernier*
Principle: ten divisions of the moveable vernier scale are drawn to cover nine divisions of the main scale with the vernier in the position shown, zero coincides with 12 and 10 coincides with 21 on the main scale, thus the reading from the right-hand end of the vermix would be 21 − 9 = 12, because the vernier covers nine divisions of the main scale

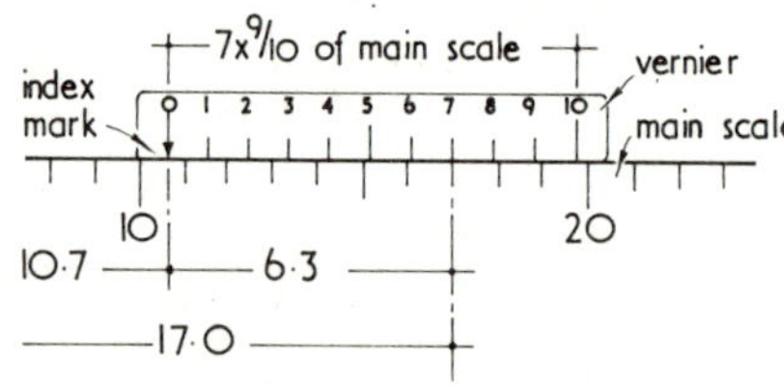

3 *Direct Vernier*
In this example the vernier division 7 coincides with the main scale division 17. Therefore to find the point where the index mark of the vernier falls it is necessary to subtract a length of 7×$\frac{9}{10}$ from 17, i.e. 17 − 6·3 = 10·7

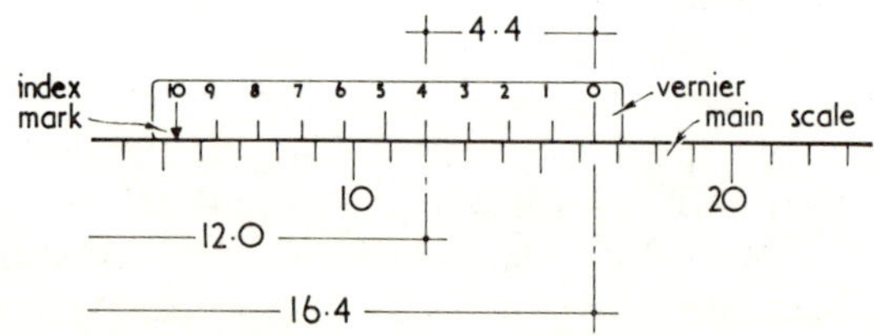

4 *Retrograde vernier*

incidence occurs at 12 on the main scale and 4 on the vernier. The reading is thus 12 + (4 × 1·1) = 16·4

As on direct verniers, the same result can be obtained more easily by counting back from the nearest whole number of the main scale (16) to the point of coincidence on the vernier scale (4).

Surveying instrument verniers

These are almost invariably concerned with angular rather than decimal measurements. Some typical examples are shown in **5**, **6**, **7**. The principle is the same but in the first example, which represents a direct vernier, since the two scales run the same way, thirty vernier divisions correspond with twenty-nine main scale divisions. However, the main scale is divided into 30min *intervals* thus twenty-nine divisions = 14deg 30min and the main scale must be read to the nearest 30min before applying the vernier addition (14 as shown). In **6** the vernier is divided into sixty divisions (thirty plus halves) which correspond with fifty-nine main scale ($\frac{1}{2}$deg) divisions. Thus each vernier division represents $\frac{1}{60} \times \frac{1}{2}$deg = 30sec.

Double verniers **7** and doubly graduated main scales are sometimes used. In such cases the principle remains the same but care is necessary to

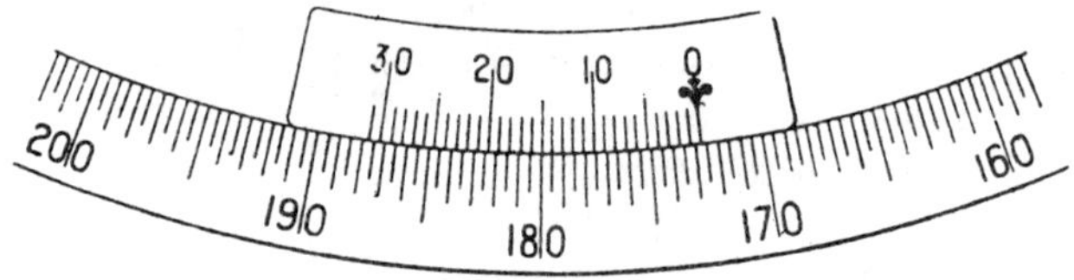

5 *Angular vernier*

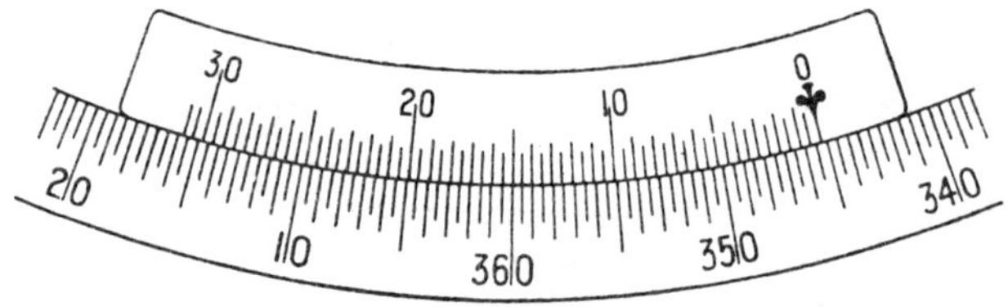

6 *Angular vernier*

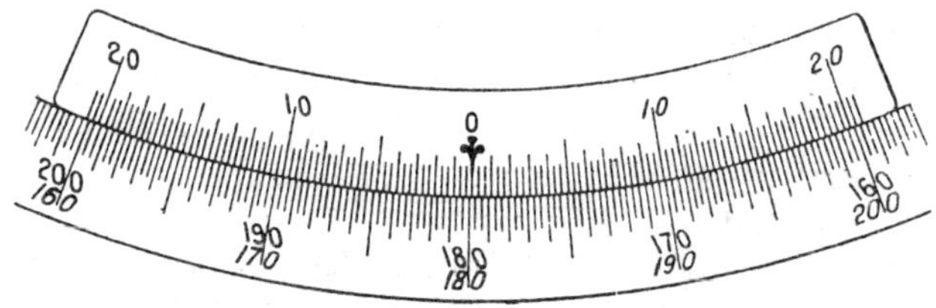

7 *Double angular vernier*

make sure that the appropriate part of the vernier is used. A useful check in such instances is that the two possible readings clockwise and anti-clockwise must add up to 360deg.

The commonest mistakes made in reading vernier scales are:

1 Failure to distinguish between a direct and retrograde vernier.

2 Failure to include subdivisions (usually 30min or 20min and 40min) before adding the vernier fractions to the main scale reading.

3 Reading the main scale in the wrong direction.

Indirect systems of angular reading

Indirect systems for reading angular scales are a feature of most modern instruments. Graduations are engraved on glass circles and light reflected through them by various means enables them to be read through a microscope eyepiece—the whole arrangement being contained within the body of the instrument. Some extremely good and effective systems are in use and make the instruments very much easier to use.

Indirect systems may include vernier scales but are more likely to have a simpler subdivision system, made feasible by good magnification and high illumination. **8** illustrates a fairly typical example of the view seen through the reading microscope of a Wild level fitted with a horizontal circle with 10min subdivisions and **9** a Watts theodolite with vertical and horizontal circle seen simultaneously.

Indirect and by micrometer

As the need for fineness of reading increases so does the vernier become less useful. One means of improvement which has become more and more widely used is the micrometer or measuring micro-

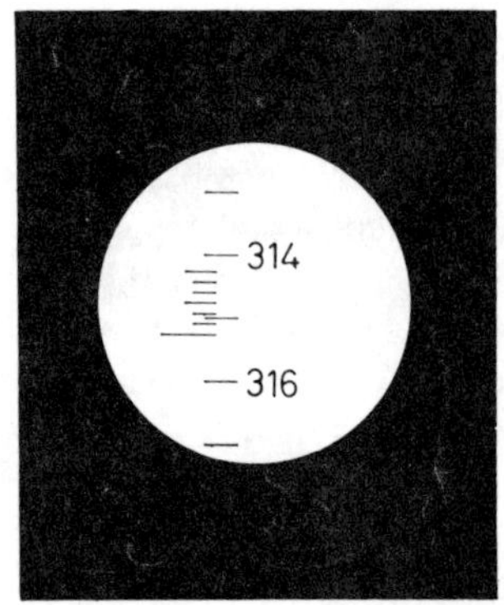

8 *Indirect optical angular scale with moving scale index fitted to Wild instruments*

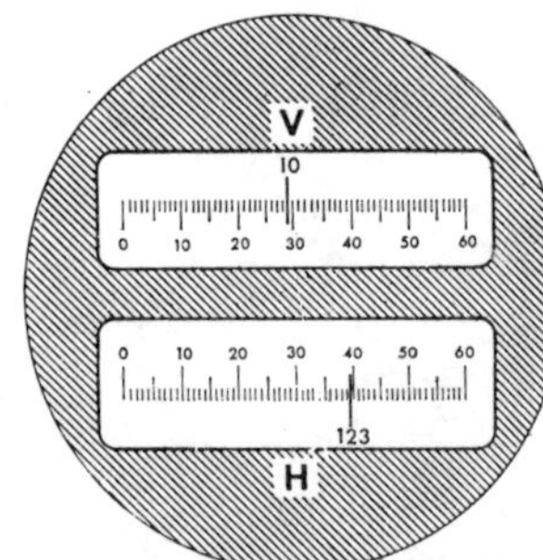

9 *Indirect angular scale with moving scale index as fitted to Hilger and Watts instruments*

scope. This is, essentially, a screw which moves the *index mark* against a scale and records the movement by a circular scale on the head of the screw. **10** illustrates the principle. In practice more refined means are used and the micrometer measures parts of the real image of the main scale (circle) divisions formed by the objective of the scale-viewing microscope. A typical example of appearance through the eyepiece is given in **11** with the micrometer before and after adjustment.

Some manufacturers combine both the horizontal and vertical circle reading into one microscope image. An example is the Zeiss Th 3 illustrated in **12**. Two micrometers are provided and the vertical scale (V) is also coloured, with its companion micrometer aperture, to prevent confusion. The micrometer may be kept at zero for quick work since angles can be estimated to 1min on the main scale without it.

Another example is the Cooke Tavistock illustrated in **13**. In this the modern tendency towards simplicity is achieved by a control which brings *either* the vertical *or* the horizontal circle into view while the same micrometer screw is used for both. An earlier version which is likely to be encountered is illustrated in **14**. The value of subdivisions (usually 20min) can be found by counting the graduations between each whole number division on the main scale.

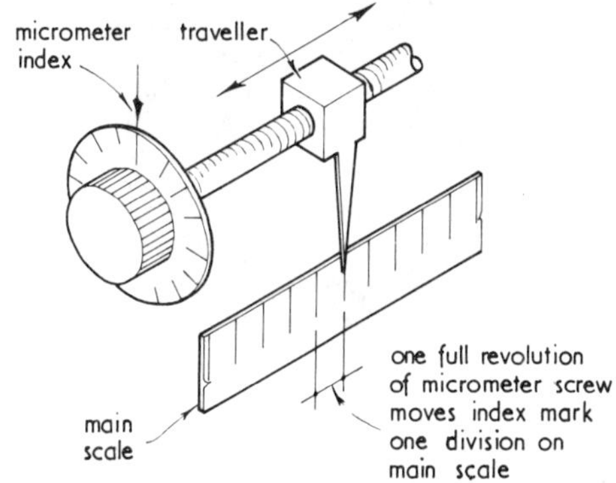

10 *General principle of micrometer indices for reading angular scales*

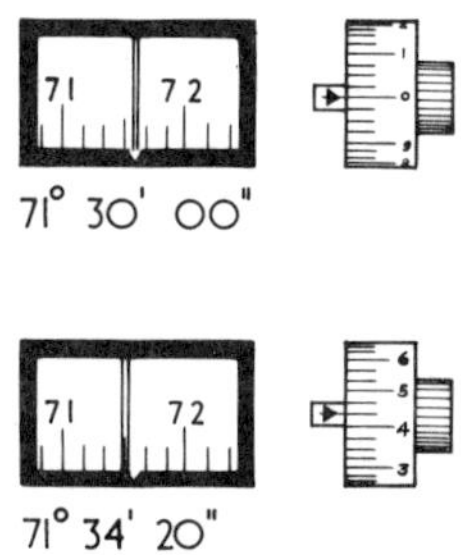

11 *Examples of angular readings with a micrometer*

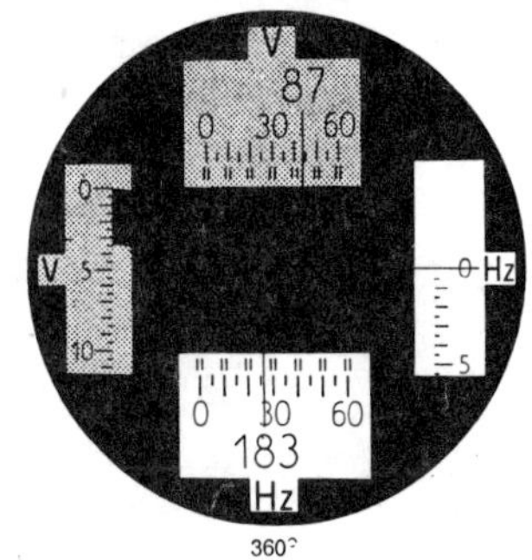

12 *Combined view of vertical and horizontal angular scales in one eyepiece as in the Zeiss Th 3*

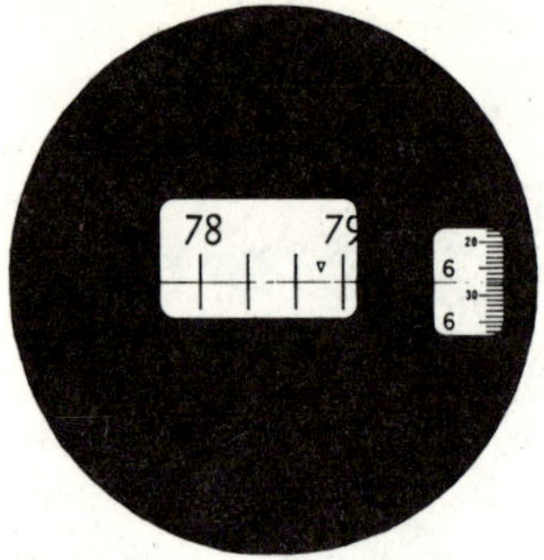

13 *View of the angular scale and micrometer reading in the same eyepiece as fitted in the Cooke (Vickers) Tavistock theodolite*

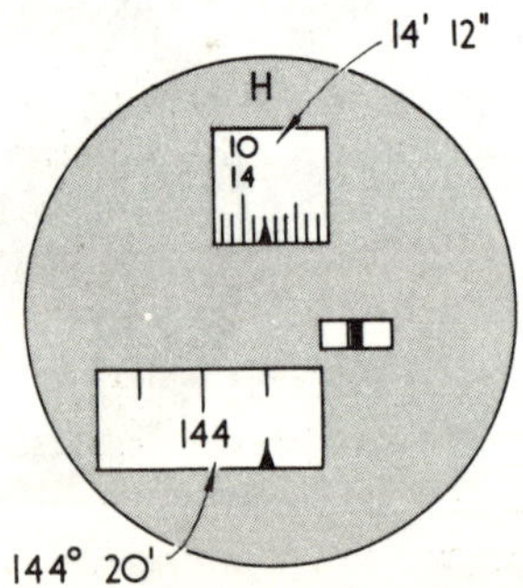

 14 *Another combined angular scale and micrometer reading system*

Chapter 4

Surveying instruments: Linear measurement by chaining

This chapter describes some of the checks and precautions that have to be taken in connection with linear measurement. It provides background information for Chapters 9 *to* 12 *which deal with techniques of surveying*

Adjustments and corrections

Surveying accuracy within the limits of capability of the instruments used depends very largely upon their state of adjustment. Some adjustments have to be made every time an instrument is used—these are called *temporary*. Other adjustments are called *permanent* because they are not often necessary.

When permanent adjustments to optical instruments have to be made it is usually as a result of wear in moving parts, temperature movement, accident or tampering. Permanent adjustments should be checked periodically but they should be made only by a skilled surveyor or instrument maker. The best procedure is to send instruments back to the maker for a periodic overhaul or if their accuracy is in doubt.

This chapter is confined to general descriptions of temporary adjustments including setting up, usually necessary in the field, and some checks of permanent adjustment and corrections that are sometimes necessary.

Ranging rods

Set up ranging rods at principal station points before beginning a survey. They should be firmly planted in the ground, or stands should be used on hard ground or paving. Check for verticality in two directions at right angles to each other using a plumb-line or mechanics level. This is especially important on sloping or irregular land.

Land chains, metallic and non-metallic tapes

Check against a steel tape before and after survey, or morning and afternoon of each day.

Steel bands and tapes

These must be handled with care to avoid forming twists or loops when laid out on the ground because they snap easily if pulled tight with a loop in them. They do not need any temporary adjustments except when a high degree of accuracy is required. Extremely accurate measurement is beyond the scope of this book so, when necessary, reference books should be consulted. The principal sources of error which do need correction are sag (catenary), incorrect tension (bands and tapes are standard only under a certain pull), slope and temperature. The latter in particular can involve otherwise puzzling differences of about 50mm in 100m for a difference of 15°c from that at which a tape or band was standardised. For normal good accuracy it is sufficient to make corrections for temperature, sag and slope.

Temperature

Temperature should be observed and recorded at hourly intervals during the period when the tape or band is in use. It is then sufficient to take an average of the observed temperatures and apply the formula: $(t_a - t_s)\ 0{\cdot}000\,021L$ where t_a = average temperature in °c, t_s = standardisation temperature (F), L = nominal length of tape or band. The correction is added if t_a is greater than t_s or, conversely, subtracted.

Sag

Steel bands and tapes are standardised for length either fully supported or 'in catenary'. The latter are meant for a special method of linear measurement and should not be purchased for ordinary architectural practice. However it is important to know which method of standardisation was used for bands and tapes already in stock, since catenary tapes will be too long if used on the ground. Correction for sag should be applied whenever it becomes necessary to make measurements with the band or tape clear of the ground (eg on sloping ground). The general condition is illustrated in **1**.

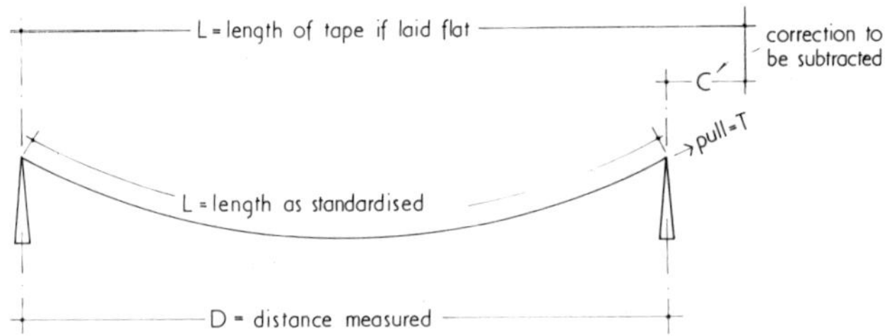

1 *Sag in catenary*

$$C = \frac{W^2L^3}{24T^2}$$ where C = correction for sag

W = weight of tape or band in kg/m

L = standardised length of tape

T = pull applied in kg (measured by spring balance)

For tapes standardised 'fully supported' the correction can be worked out in advance, provided that the same pull (T) is applied whenever the tape or band is used with a sag (in catenary). Where only part of the tape sags (eg at the end of a chain line), the correction is proportional, ie the correction is multiplied by

$$\frac{\text{length read on tape}}{\text{length as standardised.}}$$

When a tape has been standardised in catenary (ie measures the correct length when it is suspended) the correction is added to distance (D), but the field method used and corrections for partial lengths of tape are beyond the scope of this Information Sheet and text-books should be consulted.

Slope

The problem of measuring up or down slopes with chains or steel bands or tapes can be met by stretching the chain or tape between the two points concerned so that it is as nearly level as possible. This normally has to be judged by eye and invariably introduces an element of sag into the measurement. The point on the ground below the raised end of the tape is found by holding a plumb-line against the chain or tape handle and marking where it comes.

Chapter 5

Surveying instruments: Procedure in the field

This chapter describes the ways in which instruments are set up and used for observation in the field. It provides background material for Chapters 9–12.

Plumb-bobs and spirit levels

Plumb-bobs and spirit levels are employed mainly for setting up and as such their use is covered elsewhere. However, preparation for setting up should cover marking station points over which instruments can be centred.

Station points

In soft ground ranging poles leave a hole which is an adequate mark for chaining measurements and centring levels.

For more accurate work in soft ground, wooden pegs with a marker nail or panel pin at the top or lengths of galvanised tubing (about 60cm) can be driven in.

For permanent marks these are inadequate and may cause damage to farming implements, thus when identification for months or years is needed reference should be made back to a wall or some similar permanent feature as shown in the following diagram.

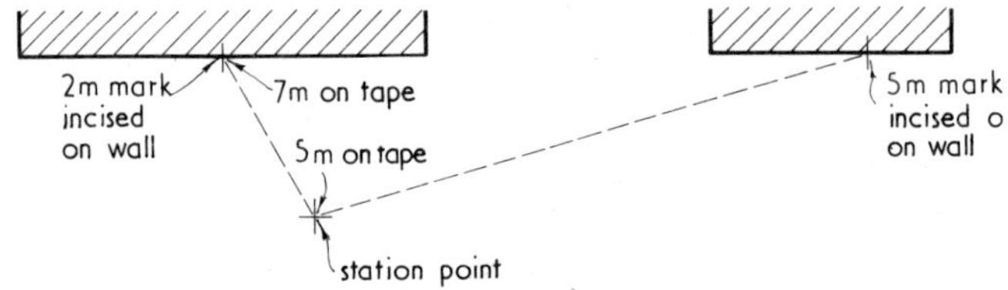

1 *Identification of station points by measurement: the distance from the wall mark to the station point is marked on the wall*

On hard standings a paint mark (red oxide paint) will usually suffice.

On hard standings, roads etc, for more accurate work a surveyor's nail or a stout panel pin should be driven and ringed with paint for short term marking, but measured identification, as described in **1** is advisable for permanency.

Chaining

Chains are normally stored and transported folded in successive pairs of links from the centre mark and strapped into a bundle. This is important since hours can be wasted in attempting to untangle a chain. To begin work unstrap the bundle, holding

it in the right hand. Take hold of the handles with the left hand and fling the bundle away from you with the right. Then walk with one handle held so as to extend the chain to its full length.
To measure with a chain one man holds the handle against the station mark **2** while a second pulls the chain out into line with the next station point. When it is straight and correct in direction he inserts an 'arrow' into the ground (or makes a chalk mark), just touching the point of the handle, signals and walks on.

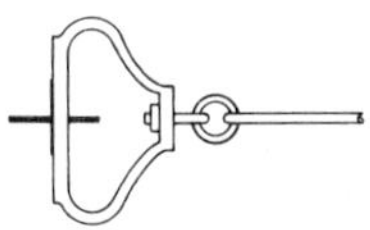

2 *Handle of chain which should be held against station mark*

When the man at the other end of the chain reaches the arrow he signals the other to stop, holds his handle against the arrow and after the forward man has put in his second arrow takes up the first and carries it with him and so on until the objective station point is reached.
This procedure is designed so that the number of arrows held by the rear chain man is a tally of the number of chain lengths. The number of chain lengths plus the portion of chain length between the last arrow and the station point is the length of the chain line.

Bands and steel tapes

A similar procedure is followed with bands and tapes except in the rare cases when a high degree of accuracy is required. On such occasions pegs or a section of board pegged down on which a fine line can be drawn must be substituted for arrows. More refined methods are beyond the scope of this book.

Tapes and ranging rods

Metallic and non-metallic linen tapes will normally be used to measure perpendicular offsets from a chain line while the chain (or band or steel tape) is still in position (at each chain length of the survey). For short distances up to about 3m it is usually sufficient to estimate right angles from the chain line by eye. For slightly more accurate work successive triangulation is more convenient. This is illustrated in **3**.

Optical squares

When long offsets are required it is convenient to use an optical square of some kind according to the degree of accuracy required. The location of the optical square on the chain line must be recorded

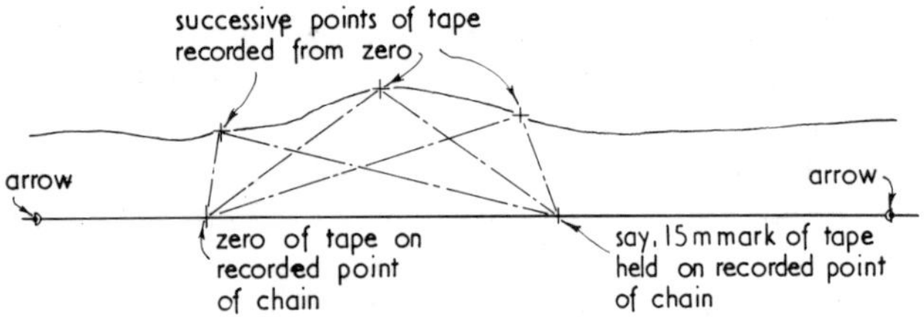

3 *Offsets by triangulation*

and a ranging rod set up to mark the point of the offset before measurement.

Optical instruments (tacheometry)

Tacheometry by stadia lines can be done with most modern levels, theodolites and tacheometers. The procedure for setting-up is as described in Chapter 7.

The most usual forms of reticle marking for stadia readings is shown in **4** and of these the commonest form is (*a*).

The instrument is focused onto a staff held at the station point under observation. The staff must be vertical and a plumb-line should be used or better, a circular bubble fixed to the back of the staff. The staff may be a levelling staff or a special stadia staff. (Staff graduation readings are described under 'Levels' in Chapter 7).

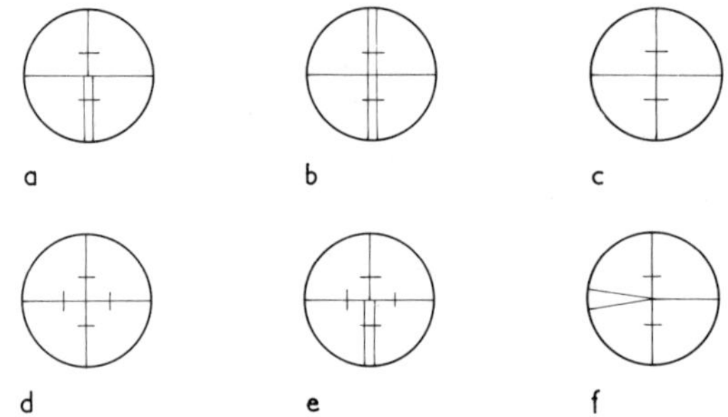

4 *Typical reticle markings for stadia reading*

The two readings intercepted by the outer pair of reticle cross lines, hairs or points are subtracted and multiplied by (usually) 100, plus or minus any correction noted in the instrument, to give the distance if the line of collimation (sight) is horizontal; or the distance which must be corrected for elevation or depression if it is not. The level reading should be made at the same time and, to check, be equal to half the sum of the stadia readings.

For tacheometrical observations in which a vertical angle is involved and corrections have to be made, as distinct from self-reducing instruments which do it for you, time and effort can be saved by using tacheometric tables such as those by Redmond (Technical Press Ltd) or Louis and Caunt (Edward Arnold Ltd).

Beaman arc

Some instruments are fitted with a Beaman arc to simplify calculation. This is, in one form or another, two extra scales engraved in the vertical circle **5**. One gives figures for obtaining difference in height between the horizontal line of sight and the point at which the middle line of the reticle intercepts the staff at the station point. The other is to give correction required to find the horizontal (plan) distance from the instrument station point to the observed station point. (See Chapter 2.2)

The procedure for making corrections is as follows: Point the telescope to read anywhere on the staff and bring the altitude bubble to zero. Then move the telescope up or down until the arc index line comes exactly opposite the nearest whole number on the V scale. Read the stadia intercepts and centre intercept. Then

(*a*) the stadia intercept multiplied by the whole number of the V scale gives the difference in height (see **5**).

Height difference = stadia × V scale.

(*b*) The stadia intercept multiplied by 100 minus stadia intercept multiplied by the reading in the D scale gives the horizontal distance.

Distance = 100 stadia—(stadia reading × D scale).

Self-reducing tacheometers

Several modern instruments are fitted with automatically variable stadia intercepts. These vary

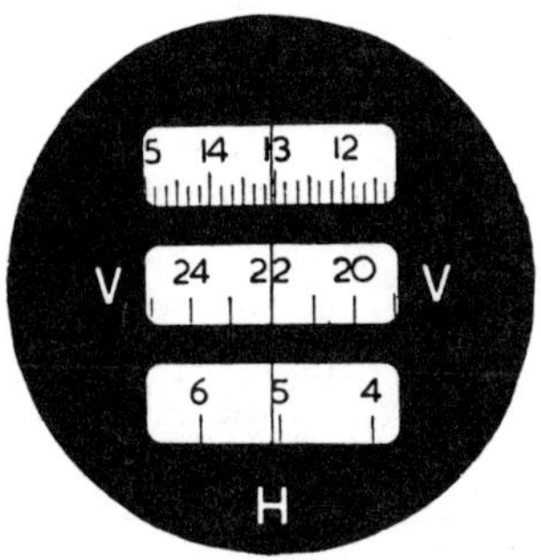

5 *Beaman arc scales seen in an indirect optical system eyepiece*

6 *Automatically variable stadia intercepts as fitted in the Wild RK 1*

according to the angle of elevation and appear in the eyepiece as sloping lines **6**.

In instruments of this kind the difference in reading between the two outer curves appearing in the reticle gives the distance directly after multiplication by 100. The vertical height is given by the reading (intercept) of the intermediate curve after multiplication by 100 and a constant marked on the curve. + or − placed before the constant indicates elevation (+) or depression (−). The vertical angle is given by a separate index.

Double image or precision telemeters or wedge attachments for theodolites

These instruments are set up in the normal way but the staff at the objective station point (target) has to be set up horizontally on a special stand. These stands usually incorporate an arrangement for sighting back to the tacheometer and a means for fixing the horizontal stadia rod at a fixed height above the ground. The graduation and figures are put upside down so that they appear the right way up in the instrument. It must be remembered that the distance obtained by reading the stadia rod is that of the rod itself; therefore the rod must be plumb over the station point and not tilted.

The stadia rod is sighted through the telescope of the instrument. The image seen will be that of a continuous run of numbered graduations with a vernier below. Most instruments have a micrometer screw which enables the vernier image to be set so that that of its graduations which most nearly corresponds with a mark in the upper (continuous) scale can be brought into exact correspondence. When this has been done the vernier can be read. The micrometer reading is then added in as shown in **7**.

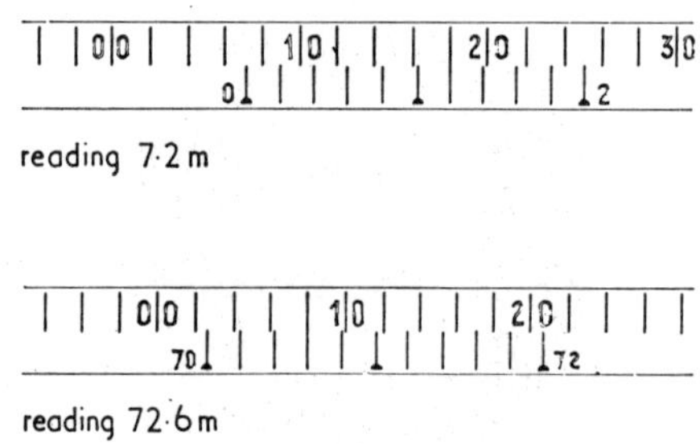

7 *Stadia rod verniers seen through a double-image telemeter eyepiece*

Chapter 6

Surveying instruments: Measurement of angles

This chapter describes temporary adjustments, checks and observational procedure for instruments primarily intended for measuring angles.

GENERAL

There are various methods for measuring angles and differences in height. This chapter describes those methods of use to an architect. It covers the temporary adjustments checks, setting-up and observational procedure employed for the more important instruments described in Chapters 2.4, 2.5, 2.6 and 2.7.

Setting up a plane table

Set up tripod, with its feet pushed into ground if possible, so that head is approximately level. Attach board and paper, plan or map.

Level board with circular bubble spirit level either by adjusting legs or loosening outer screw of ball and socket mounting and moving gently until level. Tighten screw and check with tube spirit level, rotating board through 360deg. Repeat as necessary.

Mark a point on the paper to represent the plane table's location, put pointer of plumbing fork on point and mark ground point. (A plumbing fork is a U shaped instrument like the frame of a fretsaw, used in conjunction with a plumb line to check that a point on the ground is immediately below a point drawn on the plane table to represent it.)

Alternatively, when taking up a second position or using a plan or map, the plumbing fork should be used when setting up (6.2.1 and 6.2.2) to locate the plan point on the board over the point on the ground that it represents. This requires a good deal of adjustment and readjustment but extreme accuracy is seldom necessary in practice.

Adjustments to a telescopic alidade are similar to those for a level or theodolite. They are dealt with in paras 6.5 to 6.5.6.

Using a compass

Trough and sight compasses should be rested on a firm surface or plane table before operating the knob or lever that frees the compass needle.

Prismatic compasses are held by the thumb through the fold-down ring provided and put up to the eye, sighted onto the object and held while the reading is made through the prism eyepiece.

Instrumental compasses are either incorporated

into theodolites, made to fit over telescopes or their standards or made to fit tripods in place of their normal instrument. They usually have some device to free the needle (4.1) and this should be used only during observation.

Check for compass accuracy

Take a compass bearing from one station point to another between 25m and 50m away. Take a back bearing from the second station back to the first. There should be a difference of 180deg between the bearings. The proximity of ironwork or electrical equipment may be a cause of error and if there is one the check should be extended to a third station.

Optical instruments generally

Almost all optical surveying instruments have to be set up on a tripod before use and many of the temporary adjustments that follow are very similar. For this reason they are described here in general terms and merely indicated by index letters where applicable for individual types of instrument.

A Set up the tripod. Pull out lower portion of telescopic legs and tighten extension clamping screws. See that the head screws are loose. Place feet of tripod as equilaterally as possible about 50cm from the station point (marker) on the ground. On sloping ground adjust by shortening the appropriate telescopic leg or, if a solid leg tripod, put two legs at the lower level parallel to the slope and the third leg at a flatter angle. Press feet of tripod into ground as far as possible. See that the head plate is as level as possible and tighten head clamping screws. Some stands have a cap with a hole in it which facilitates centring over the ground point with a plumb-bob. Remove cap and place in box lid.

B Take out instrument and see that all tribrach and alidade clamping screws are loose, then screw instrument to tripod head so that it is firm but not excessively tight.

GENERAL WARNING *Never use force on any screw on an optical instrument*

C Attach plumb-bob. If the instrument is not approximately level it may be adjusted by repositioning two legs of the tripod simultaneously. If a circular bubble is provided the bubble should at least be touching the index ring at this stage, with the levelling screws at about the centre of their run. Check tripod head clamping screws and firmness of tripod.

D Level the instrument with the three levelling screws (the procedure is somewhat different for older four-screw models) thus:

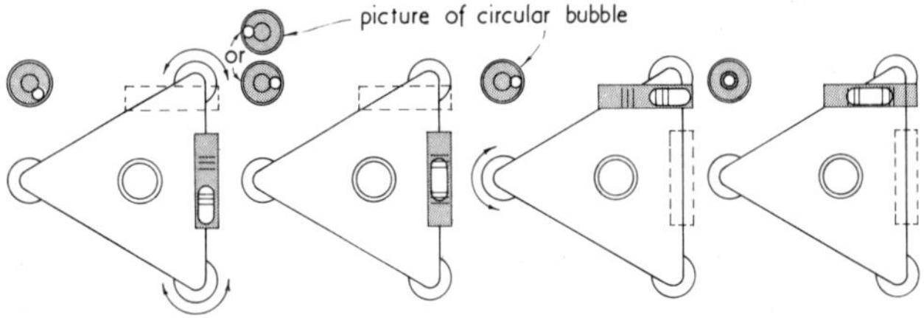

1 *Step 1 Place the plate bubble over two levelling screws.*
Step 2 Counter-rotate them simultaneously to bring the bubble to the centre of its run.
Step 3 rotate the plate through 90°.
Step 4 turn the third levelling screw to centre the bubble again

E Put the telescope objective out of focus and then focus the eyepiece—normally by rotating it—so that the reticle image is as clear and sharp as possible. This setting will vary according to each observer's eyesight. Theoretically (for a single observer) it needs doing only once during a survey, but in practice the eyepiece is sometimes put out of focus accidentally when moving from station to station.

F Focus the telescope onto the object (levelling staff, etc) and when the image is sharp turn your head slightly to and fro, sideways. The image of the object should seem to have the image of the reticle fixed to it. If the one moves relative to the other there is an error of *parallax* and the telescope must be refocused until there is none.

G Check verticality of vertical axis: After levelling place the levelling bubble across two levelling screws as in **1** (step 1) and rotate the telescope through 180deg, the bubble should remain central. Repeat check after step 2. When correction is necessary it is made, after the step 1 check, by means of the capstan-headed screws at the end of the tubular bubble. These are turned with the small tommy bar (normally provided in instrument box) to correct the bubble displacement by half its error. The telescope is then turned back again through 180deg and relevelled with the levelling foot screws. Repeat the check until correct.

Note: this is a *permanent* adjustment. It should not normally be necessary and is better done by an expert. Correction of more sophisticated instruments with coincidence bubble or automatic levelling devices should not be attempted.

Telescopic alidade

When a telescopic alidade is used, horizontality of the plane table as already described should be checked with the alidade level. Then perform steps E and F (described above).

Cowley level

Put up tripod reasonably vertically, place instrument on the pivot pin. It is then ready for use. *Check*. Select two firm positions for the target about 15m apart. With the level as close as possible to one of them measure the difference in level. Repeat with the level close to the second target point. There should be a negligible difference in results.

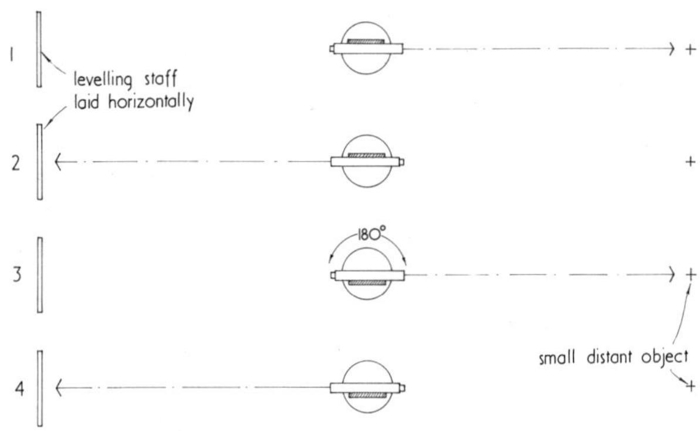

2 *Test to see that the line of collimation does pass through the centre of the cross hairs of the reticle: make the sightings shown reversing the telescope at Step 3. Fourth step should produce the same reading as the horizontal staff*

Dumpy level

Perform A, B, C, D, E, F and G.

Quickset levels and others with hemisphere and cup mounting

Perform A and B.

Loosen the clamping ring or screw so that the instrument moves freely in its socket. Move until circular bubble is central, hold the instrument steady and tighten up gently. When the clamp barely grips, adjust again so that the bubble is central and tighten cautiously so that the clamp is firm. Perform E and F. Do not attempt G.

Reversible, automatic and precise levels

Perform A, B, C, D, E and F.

Theodolites and tacheometers generally

Perform A, B, C, D, E and F.

Make the final orientation adjustment of the vertical axis over the ground point, using the sliding plate or similar arrangement on the trivet stage to do so, and clamping in position when centred. An optical plummet is often incorporated in modern instruments and, if it is, the first rough centring can be done by plumb-bob and completed with the optical plummet.

Turn the alidade so as to put the vertical circle

level tube over a pair of levelling screws and bring the bubble to the centre of its run by means of the index setting screw and repeat D. (Note, the vertical index bubble is usually so sensitive that it may be affected by the weight of the observer moving round the instrument.) It is rarely possible to adjust the levelling screws so that the bubble remains absolutely at the centre of its run when the telescope is rotated—it is sufficient if it *almost* does so—provided that it is checked before each reading of the vertical circle. To check that the line of collimation does pass through the centre of the cross-hairs in the reticle—assuming that the latter is correctly marked with the vertical and horizontal at 90deg—proceed as shown in **2**. If there is an error it should be corrected by an instrument maker.

Chapter 7
Surveying instruments: Measurement of height

This chapter describes temporary adjustments, checks and observational procedures for the level

Levelling

Once instruments have been set up and the necessary temporary adjustments completed they are ready for observations to be made. It is advisable to follow a set routine for this as for setting up etc. Observation routines may vary a little with different makes of instrument, especially if individual manufacturers have incorporated special refinements of their own. The procedures given here are intended as a guide on the general use of instruments.

Levels

Differences of level are read from a scale drawn on a staff which is put at the observed station point. Accuracy diminishes with distance as the staff image grows smaller. The magnification of the telescope therefore regulates the distance over which effective observations can be made. Some typical distances representing the limit at which it is possible to read down to 5mm are given in the table below.

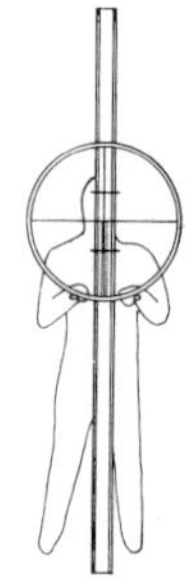

1 *Holding the levelling staff*

Maximum distance at which 5mm interval can be distinguished

Magnification	Metres
18 ×	200
20 ×	230
24 ×	280
28 ×	360
32 ×	400
42 ×	450

Note Some instruments have a resolving power exceeding these figures.

The image of the levelling staff is seen upside down in some instruments: divisions therefore have to be reckoned upwards. **2** illustrates a typical reading at mid-distance.

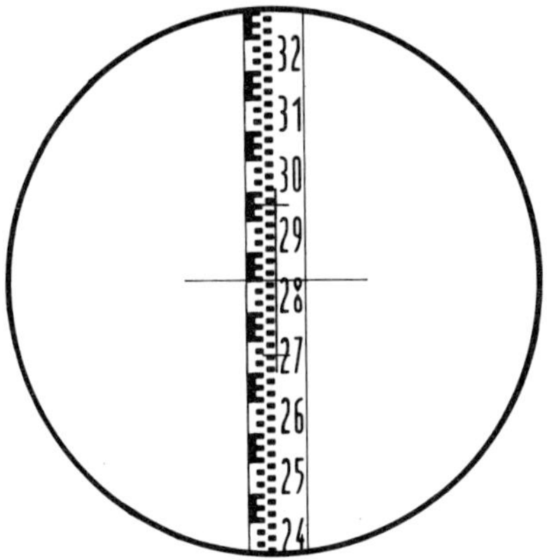

2 *The levelling staff as seen through an instrument*

Procedure The instrument is set up as previously described, then the levelling staff is put with its base on the point where level is to be measured by a 'staff man': the staff, extended only as much as may be necessary, should be held facing the level by placing the two hands flat against its sides, at about chin level, with the feet well apart at an equal distance each side of the staff base. Unless a circular bubble is provided the staff should be moved slowly forward and backward so that the observer with the level can estimate when it is vertical (ie the point at which the greatest reading is obtained). The observer can correct for sideways tilt, which is obvious from reference to the reticle's vertical lines, by signalling the staff man to move the top to the right or left. The staff is read where the central horizontal line of the reticle intercepts the staff graduations. Readings should be booked. (See Chapter 12) and checked before the staff man moves to his next position.

Measuring horizontal angles

The method employed to observe horizontal angles varies according to the type of instrument and the degree of accuracy required.

Plane table When a plane table has been set up and is in adjustment survey lines can be drawn from the point on the paper representing the station point at which it is set up by sighting on objects with the alidade. Most alidades have a parallel-rule attachment and the method is to place the alidade with the straight-edge against, or nearly against, the station point, sight onto the object, move the parallel rule against the station point and draw a line. When this is repeated for a number of objects the angles between them will be recorded graphically.

Linear measurements from station point to station point along survey lines are marked off, to scale, from dimensions obtained by chaining or tacheometry.

When all necessary observations have been made at one station point the plane table is taken to the next and the procedure repeated.

Detail is filled in as work proceeds, or after the main framework of chain lines has been drawn. It is done, to scale, either from normal chaining measurements with offsets or by tacheometrical observations for distance and sightings for direction.

Levels The following describes levels and other instruments with a simple horizontal graduated circle, where there is no separate provision for 'zeroing' the horizontal circle.

1.1 The telescope is sighted approximately onto the objective (target) by means of the 'gun-sight' fixed to the barrel. If this is not done it is difficult to find the object through the telescope.

1.2 When the object image appears against the reticle it is brought into focus.

1.3 The asimuth clamp is tightened and the slow motion screw used for final adjustment to bring the object to the centre of the reticle, moving left to right.

1.4 The level-bubble is checked and if necessary adjusted to the centre of its run.

1.5 Level or stadia readings are taken and recorded (See Chapter 12)

1.6 The horizontal angle is read and recorded (See Chapters 3 and 11)

1.7 The object image is displaced to the right and brought back to the centre of the reticle from right to left with the slow motion screw. The horizontal angle is again read and if it differs, the average of readings 1.6 and 1.7 is taken.

1.8 The level reading is checked before signalling the staff man to move.

1.9 Steps 1.1 to 1.8 are repeated for subsequent objects.

Theodolites Theodolites, some tacheometers and similar instruments are fitted with two clamps and slow motion screws controlling the upper and lower plates of the alidade. Procedure when taking horizontal angles is: 2.1 Rotate the upper plate against the lower plate until the index mark for the horizontal circle is at zero (360deg on some instruments) and 180deg if both sides of the circle are readable. Tighten the upper plate clamp and use the slow motion screw for fine adjustment.

2.2 Check that the micrometer screw (if there is one) is at zero. Also check that one rotation moves the index over the appropriate pair of graduations. If it does not it will be necessary to read angles both forward (normal) and backwards from the higher graduation and take an average by $\frac{r_1}{d}(r_2 - r_1)$ where where d = value of one revolution r_1 = normal reading and r_2 = backward reading.

2.3 With the two plates clamped together (2.1) sight on to the first object (target) using the gun-sight for a first approximation.

2.4 Tighten the lower plate clamp and bring the object onto the centre line of the reticle, from left to right, with the slow motion screw.

2.5 Loosen the upper plate clamp and swing the telescope to the second object, clockwise, and focus.

2.6 Tighten the upper plate clamp and bring the object to the centre of the reticle using the *upper* plate slow motion screw for final adjustment from left to right.

2.7 Read the angle and record it. (See Chapter 11)

2.8 Repeat 2.5 to 2.7 for any other objects to which angles are to be read.

2.9 Repeat the whole procedure again but with the telescope reversed and in an anti-clockwise direction, thus a mean of two angles for every station will be obtained.

Note If the instrument is equipped with a micrometer it need not be based for fine readings if only a moderate degree of accuracy is required. The above procedure will suffice.

A higher degree of accuracy can be attained either by using a micrometer theodolite and repeating the procedure given above. Otherwise, using vernier or less sophisticated instruments, the procedure may be elaborated as follows:

3.1 Proceed as 2.1 to 2.7.

3.2 Loosen the *lower* plate clamp, aim at the station, bring the object to the centre of the reticle approximately and tighten the lower plate clamp.

3.3 Bring the object to the centre of the reticle with the lower slow motion screw. The angle previously recorded should still appear on the horizontal scale.

3.4 Repeat 2.5, 2.6 and 2.7. The angle recorded should now be double the previous reading. If there is more than a small difference from this then an error has been made and the surveyor should begin again from 2.1.

3.5 If there is no more than a small difference continue with the round of angles as 2.8 deducting the initial reading from subsequent angles.

3.6 Repeat anti-clockwise as 2.9.

Note At the conclusion of each round it is advisable to take the angle to magnetic north.

Measuring vertical angles

Theodolites, tacheometers and telescopic alidades (plane table) are the principal instruments used for taking vertical angles. The procedure is very similar for each. The essential requirement is careful attention to the level tube associated with the telescope. This usually is the most sensitive one on the instrument. General procedure is as follows:

4.1 Centre the vertical circle level tube and check by reversal through 180deg horizontally for gross error.

4.2 Zero or set the vertical circle index. Various arrangements of the graduations are given in **3**.

4.3 Sight upon the object to intercept the point to which the angle is to refer (if readings are being taken from more than one station point it is important that the point on the object, to which sights are made should be visible and identifiable from each) with the gun-sight.

4.4 Tighten the vertical circle clamp and, using the slow motion screw, intercept the level point of the object with the horizontal centre line of the reticle.

4.5 Read the vertical circle angle. The arrangements for doing so will be similar to those for the horizontal circle. If two verniers are provided, diametrically opposed, read both and record them. Note carefully whether the sight is one of elevation or depression.

4.6 Transit the telescope (reverse through 180deg vertically) and rotate the alidade through 180deg. Check the level-bubble.

4.7 Repeat the procedure from 4.2 to 4.5.

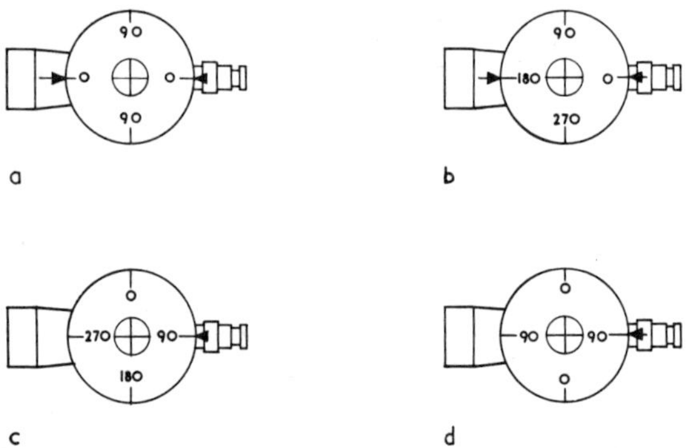

3 *Various arrangements of graduations on the vertical circles of theodolite*

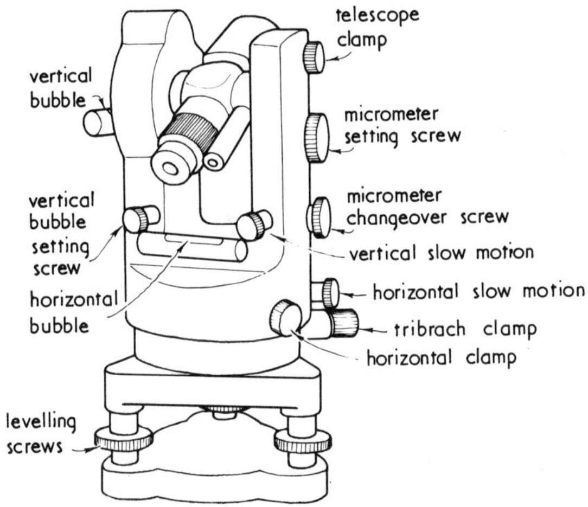

4 *Main adjustment controls of theodolite*

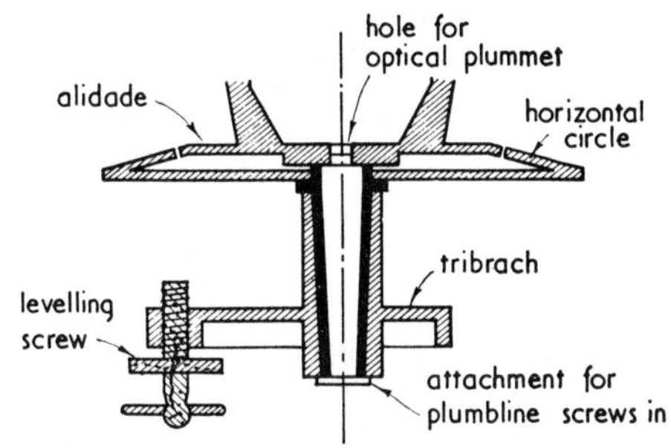

5 *Theodolite horizontal circle and its mounting*

Chapter 8

Surveying instruments: Contemporary optical instruments, manufacturers and costs

This chapter is a review of proprietary surveying instruments currently available and a guide to their cost. It deals with levels, theodolites and various specialised instruments

General

Surveying instruments have undergone many refinements in design and manufacture over the last twenty years. The general trend has been towards greater reliability, ease in manipulation, more robust construction and enclosure of parts that are easily damaged. Tables I, II and III list types of proprietary equipment, their characteristics and approximate cost. The list of manufacturers is representative rather than exhaustive.

Overhaul and repair of instruments

Some firms have a service for overhauling instruments (including making permanent adjustments) and will undertake to bring instruments into good serviceable condition. As an example of the cost, tables IV and V list rates charged by Survey and General Instrument Co Ltd for any of the makes of instruments mentioned. Lower charges are made for checking and cleaning which does not involve stripping the instrument down. Facilities offered include a collection and delivery service (within a specified radius of the firm's works), insurance during servicing and repair, and special rates for contract servicing.

Major repair work is additional to the above, the charge depending on the nature and extent of the repairs.

Hire of instruments

Instruments can be hired by the day, week or month. Alternatively, it is possible to rent instruments (as distinct from hiring them) which means entering into a renewable contract for a specified period (one, two, three years and so on). Examples of rates charged by Survey and General Instrument Co Ltd are given in tables VI and VII.

Table I *Contemporary optical surveying instruments: Levels*

Manufacturer	Maker's description of instrument		Telescope magnification	Minimum focusing distance	Greatest distance to read 1cm	Stadia ratio	Horizontal circle graduation interval
Fuji	Builder's automatic level	AL31	20×	Not stated	200m	No stadia lines	1deg
Hilger & Watts	Quickset tilting	SL10	14×	3m	210m	1 : 100	1deg
	Engineer's tilting	SL130	25×	1·8m	270m	1 : 100	10min
	Engineer's automatic	SL70	25×	2·6m	255m	1 : 100	1deg
	Precise automatic	SL62	32×	1·8m	300m	1 : 100	10mm
Kern & Co	Simple builder's level	GK0C	18×	0·9m	250m	1 : 100	1deg
	Engineer's small level	GK1C	22·5×	0·9m	300m	1 : 100	1deg
	Engineer's automatic level	GK1AC	25×	2·3m	320m	1 : 100	1deg
	Engineer's precise level	GK23C	30×	1·8m	400m	1 : 100	1deg
Vickers Instruments (previously Cooke, Troughton & Simms)	Vicker's small automatic	S27	21×	1·8m	Not stated	1 : 100	2deg
	Cooke surveyor's level	S3370	28×	2m	340m	1 : 100	10min
	Cooke engineer's level	S440	32×	3m	400m	1 : 100	5min
	Cooke self-aligning level	S77	32×	1·8m	360m	1 : 100	5min
Wild Heerbrugg	Dumpy level	NK01	19×	1·6m	200m	1 : 100	1deg
	Engineer's compact level	NK10	20×	1m	200m	1 : 100	1deg
	Engineer's level	NK2	24×	2m	270m	1 : 100	1deg
	Precision level	N3	42×	2·15m	400m	1 : 100	—

Method for reading horizontal circle and minimum reading		Tilting screw gradienter or p.p. micrometer	Method of reading level bubble	Bubble sensitivity per 2mm	Levelling accuracy per km	Notes	Approximate cost plus tripod
Direct	0·1deg	—	Automatic	—	Not stated	Horizontal circle can be rotated for zero setting	£114
Direct	Not stated	Tilting screw	Direct	Not stated	8mm	A long service model	£68
Direct	5min	Tilting screw	Coincidence	Not stated	2mm	Erect image with SL130,	£150
Direct	Not stated	—	Automatic	Not stated	2mm	SL70 and SL62	£127
Direct	5min	—	Automatic	Not stated	1·2mm		£214
Direct	6min	Tilting screw	Direct	40sec to 50sec	7mm	Kern levels have no levelling	£79
Direct	6min	Tilting screw	Coincidence	40sec to 50sec	2·5-4mm	screws due to patent head.	£113
Direct	6min	—	Automatic	$\pm$0·5sec	2·5mm	Prices about £7 less if without	£186
Optical	1min	pp micrometer	Coincidence in telescope eyepiece	18sec	2mm	horizontal circles	£207
Direct	Not stated	—	Automatic	—	Not stated	The horizontal circles of levels	£106
Optical	1min	Tilting screw	Coincidence	50sec	Not stated	S3370 and S440 can be rotated	£139
Optical	1min	Gradienter	Coincidence	25sec	Not stated	independently for zero setting	£182
Optical	1min	pp micrometer	Automatic	Not stated	Not stated		£199
Direct	6min	—	Direct	60sec	5mm	The NK10 and N2 levels	£79(A)
Optical	1min	Tilting screw	Coincidence	60sec	2·5mm	can be supplied without a	£129(A)
Optical	1min	Tilting screw	Coincidence	30sec	2mm	horizontal circle	£160(A)
—	—	Tilting screw and pp micrometer	Coincidence	10sec	0·2mm		£295(N)

Some instruments have non-telescopic tripods, these are marked (N) against the cost figure. Some may have either telescopic or non telescopic tripods, these are marked (A).
The costs given are approximate.

Table II *Contemporary optical surveying instruments: Theodolites*

Manufacturer	Maker's description of instrument		Telescope magnification	Minimum focusing distance	Greatest distance to read 1cm	Stadia ratio	Methods for reading level bubbles	Bubble sensitivity per 2mm Plate	Vertical circle
Hilger & Watts	1 sec microptic theodolite	ST200	28×	1·8m	450m	1 : 100	Direct and coincidental	20sec	20sec
	10sec auto including microptic theodolite	ST302	28×	1·8m	450m	1 : 100	Direct	30sec	Automatic
	20sec microptic theodolite	ST156	25×	1·65m	450m	1 : 100	Direct	30sec	30sec
W. F. Stanley & Co	Builder-contractor	A68	14×	Not stated	Not stated	—	Direct	1min	1min
	Two-micrometer	A25	24×	6½ft	Not stated	1 : 100	Direct	30sec/1·6mm	20sec/1·6mm
	Eltham	A21	24×	6½ft	Not stated	1 : 100	Direct	30sec/1·6mm	20sec/1·6mm
	A1 micrometer		20×	4ft	Not stated	1 : 100	Direct	1min/1·6mm	30sec/1·6mm
Vickers Instruments (previously Cooke, Troughton & Simms)	Cooke	V11	15×	Not stated	Not stated	1 : 100	Direct	Not stated	Not stated
	Tavistock	V406	25×	1·8m	Not stated	1 : 100	Direct and coincidental	20sec	20sec
	Cooke	V22	25×	1·8m	Not stated	1 : 100	Direct and coincidental	45sec	90sec
Wild Heerbrugg	Compass theodolite	T0	20×	1·4m	220m	1 : 50 & 1 : 100	Direct	8min	8min

Method for reading circles Horizontal	Vertical	Minimum angles that can be read Horizontal	Vertical	Instrumental averaging angles	Simultaneous view of both circle readings	Optical plummet fitted	Magnetic compass fitted	Weight without box	Notes	Approximate cost with tripod
Micrometer	Micrometer	1sec	1sec	Yes	No	Yes	Extra	6·3kg	Robust	£496
Micrometer	Micrometer	10sec	10sec	No	Yes	Yes	Extra	5·4kg	The ST302 and ST156 can be supplied with errect image telescopes	£360
Micrometer	Micrometer	20sec	20sec	No	Yes	Yes	Extra	5·1kg		£326
Vernier	Vernier	1min	1min	No	No	No	No	9lb	Light, suitable for simple work	Not stated
Micrometer	Vernier	10sec	20sec	No	No	Extra	Extra	$15\frac{3}{4}$lb	A four-micrometer version can be had	Not stated
Vernier	Vernier	20sec	20sec	No	No	Extra	Extra	$15\frac{1}{2}$lb	A quick-set head available for A21 and A26	Not stated
Micrometer	Micrometer	20sec	20sec	Yes	Yes	Yes	Extra	7lb	Robust, totally enclosed scales	Not stated
Vernier	Vernier	1min	1min	No	Yes	Extra	No	Not stated		£209
Micrometer	Micrometer	1sec	1sec	Yes	Yes	Yes	Yes	4·9kg		£633
Moving scale	Moving scale	10sec	10sec	No	Yes	Yes	Yes	5·2kg	Permits high-speed accuracy	£366
Direct	Direct	2min	1min	Yes	No	No	Yes	2·9kg	Horizontal circle acts as a compass on T0 model	£263
Micrometer	Micrometer	5sec	5sec	Yes	Yes	Yes	No	5kg		£390

Table II – *continued*

Manufacturer	Maker's description of instrument		Telescope magnification	Minimum focusing distance	Greatest distance to read 1cm	Stadia ratio	Methods for reading level bubbles	Bubble sensitivity per 2mm Plate	Vertical circle
	Double-centre	T1A	28 ×	1·5m	300m	1 : 100	Direct and automatic altitude	30sec	Automatic
	Direct reading theodolite	T16	28 ×	1·5m	300m	1 : 100	Direct and coincidental	30 sec	30sec
	Universal theodolite	T2	28 ×	1·5m	300m	1 : 100	Direct and coincidental	20sec	20sec
Carl Zeiss	Theodolite	TH3	25 ×	1·2m	Not stated	1 : 100	Not stated	30sec	Automatic
Kern & Co	Small engineer's DC theodolite	DKM1	20 ×	0·9m	300m	1 : 100	Direct	30sec	30sec
	First order DC theodolite	DKM2	30 ×	1·7m	410m	1 : 100	Direct	20sec	20sec
	Engineer's theodolte	KI-A	28 ×	1·8m	370m	1 : 100	Automatic setting	40sec	10sec

Instruments stated to have simultaneous viewing of both circle readings may have a device to bring horizontal circle readings and vertical circle readings into view successively in one eyepiece. Costs given are approximate. Many accessories are available at extra cost.

* Converts to self-reading tacheometer with Ewing Stadi-Altimeter. See under Miscellaneous instruments.

† Theodolites supplied with aluminium centering tripods as standard.

Method for reading circles Horizontal	Vertical	Minimum angles that can be read Horizontal	Vertical	Instrumental averaging angles	Simultaneous view of both circle readings	Optical plummet fitted	Magnetic compass fitted	Weight without box	Notes	Approximate cost with tripod
Micrometer	Micrometer	6sec	6sec	Yes	Yes	Yes	No	4·7kg		£390
Micrometer	Micrometer	1sec	1sec	Yes	No	Yes	Extra	5·6kg	T2 adequate for all forms of survey	£572
Micrometer	Micrometer	0·005min	0·005min	Yes	Yes	Not stated	Not stated	Not stated		Not stated
Moving scale	Moving scale	10sec	10sec	Yes	Yes	Extra	Extra	1·8kg	Very light, rugged and accurate	£420†
Micrometer	Micrometer	1sec	1sec	Yes	Yes	Yes	Extra	3·6kg	For high accuracy work	£580†
Micrometer	Micrometer	20sec	20sec	No	Yes	Extra	Extra	4·2kg	No vertical or horizontal clamps—reversible reading horizontal circle—automatic setting of alidade	£420†

Table III *Miscellaneous instruments*

Manufacturer	Instrument	Characteristics
Hilger & Watts	Microptic alidade	Telescope magnification 15 × stadia ratio 1 : 100 Altitude level sensitivity 60sec per 2mm Cross bubble sensitivity 2min per 2mm Blade length 375mm with parallel rule fitted Scales: vertical circle, subdivisions 10min, estimate to 1min, Beaman arc scales, vertical and horizontal Telescope can be transited Vertical axis can be adjusted independently of plane table
Hilger & Watts	Ewing stadia altimeter	A device which can be fitted to a number of this firm's theodolites to convert them into direct-reading tacheometers. It is, in effect, a small computer which can be set with stadia readings from the theodolite on which it is fitted to give difference in height and horizontal distance without reading the vertical circle
Hilger & Watts	Tacheometer prism and staff	The outfit comprises a prism unit which can be fitted to a theodolite telescope plus a horizontal staff and stand. It enables distance to be measured optically with an accuracy of up to 1 in 3000. The staff incorporates a displacement vernier graduation (see AJ Information Sheet 1346), while the prism has a micrometer which makes it possible to obtain range readings down to 1cm
Wild Heerbrugg	RDS self-reducing tacheometer	Telescope magnification 24 ×. Shortest possible range about 3·3m. The reticle incorporates automatically variable stadia lines permitting direct stadia measurements to be made. The horizontal and vertical angle can be measured with an optical micrometer. Price £614
Survey & General	BRS tape tension handle	Replaces spring balance in applying the correct tension to a tape during setting out. It has been shown to virtually eliminate errors in tape measurement. Price £9
Hilger & Watts	Watts Autoplummet	This instrument is a telescope with the line of sight turned through 90deg downwards by means of a prism. This latter can be adjusted either side of its normal verticality by means of a micrometer screw. Verticality is ensured (as in the Autoplumb) by means of an automatic compensator. The range of the instrument is from 1·2m to

Manufacturer	Instrument	Characteristics
		infinity with an accuracy of 2sec of arc. The Autoplummet can be interchanged with a theodolite on the same tripod
Wild Heerbrugg	Zenith and nadir plummet	This instrument has a 10 × telescope and a tubular level of 20sec/2mm sensitivity. The shortest focusing distance is 0·6m, with a plumbing accuracy of 1 : 30 000. A control knob switches the prism which turns the line of sight through 90deg to point upwards or downwards. The plummet can be exchanged with any of the Wild theodolites on the same tripod or used as an independent instrument. Telescope can be transited and the instrument includes vertical and horizontal circles graduated to read to 6sec both being visible in one eyepiece. Price £161
Hilger & Watts	Rangefinder theodolite	This instrument operates on the image coincidence system (AJ Information Sheet 1339) with objective lenses placed each end of a 1m baseline tube—thus making it possible to make tacheometric observations without the use of a stadia staff and staff-man. Magnification 14 ×. Minimum range 100m, maximum 10 km. The instrument is equipped to read vertical and horizontal angles to 20sec
Hilger & Watts	Autoplumb	This instrument is equipped with two telescopes for sighting upwards and downwards. Magnification : upwards 30 ×, range 500ft, minimum focus 1·8m : downwards 17 ×. Minimum focusing distance 0·9m. An automatic compensator is built into the upward telescope ensuring true verticality even when the instrument is tilted up to 15min of arc. Errors in verticality (of the target point) can be measured with a certainty of 1mm in 40m. With extra care this can be improved to 0·3mm in 40m

Table IV *Typical charges for overhauling theodolites*

Manufacturer	Type of theodolite	Charge £p
Hilger & Watts	ST2 and B4 vernier	28·75
	B50 and ST110 microptic	48·00
	ST156 and ST150 microptic	41·00
	ST200 and ST202 microptic	63·75
A. H. Hall & Bros	T5 vernier	28·75
	$3\frac{1}{2}$in micrometer	38·00
Kern & Co	K1-A automatic micrometer	51·50
	DK1 and DKM1 optical micrometer	58·50
	DK2 and DKM2 optical micrometer	63·75
	DKM3 optical micrometer	95·00
	K1-RA self-reducing tacheometer	63·75
	DK-RV self-reducing tacheometer	65·00
	DK-RT self-reducing tacheometer	95·25
W. F. Stanley & Co	A68 builder's	23·50
	A21 and U2 vernier	28·75
	A26 full micrometer	38·00
	A1 optical micrometer	58·50
Vickers Instruments (previously Cooke, Troughton & Simms)	V100 and V101 vernier	28·75
	V208 optical scale	58·50
	V22 optical scale	48·50
	V301 and V302 optical micrometer	53·25
	V406 and Mk1 Tavistock optical micrometer	62·00
Wild Heerbrugg	T0 compass	38·00
	T1 and T1A optical micrometer	58·50
	T16 optical scale	58·50
	T3 first order optical micrometer	63·75
	RDS self-reducing tacheometer	63·75
	RDH self-reducing tacheometer	95·25
	T2 optical micrometer	63·75
Zeiss Oberkochen	TH3 optical micrometer	63·75

Table V *Typical charges for overhauling levels*

Manufacturer	Type of level	Charge £p
Hilger & Watts	SL10 and SL2 builder's	7·75
	Highway ball-base engineer's	13·00
	Highway three-screw base engineer's	14·75
	SL103-6-8 microptic with mirror bubble	14·75
	SL104 microptic with mirror bubble and glass circle	17·25
	SL102 microptic with coincidence bubble	14·75
	SL100 microptic with coincidence bubble and glass circle	18·25
	SL70 autoset (engineer's)	13·00
	SL60 autoset precise	18·25
	SL62 autoset with glass circle	21·75
	B125 reversible	21·75
	SL50 precise with built-in PP micrometer	23·50
A. H. Hall & Bros	7·5 builder's	14·75
	9·5 engineer's	14·75
	precision engineer's	18·25
Kern & Co	GK0 and GK0-C builder's	7·75
	GK1 and GK1-C engineer's	9·50

Manufacturer	Type of level	Charge £p
	GK1-A and GK1-AC automatic	9·50
	GK23 precision engineer's	14·75
	GK23-C precision engineer's with glass circle	18·25
	GK0-A automatic level	9·50
W. F. Stanley & Co	A450 builder's	7·75
	A500 builder's dumpy	13·00
	A520 and A525 ball-base engineer's	9·00
	A510 and A515 three-screw engineer's	9·50
	A530 and A535 three-screw engineer's	9·50
	A505 reversible self-checking	21·75
Vickers Instruments (previously Cooke, Troughton & Simms)	S201 builder's	9·00
	S200 and S220 builder's with circle	9·50
	S301 and S331 engineer's	14·75
	S300 and S330 engineer's with glass circle	18·25
	S401 and S441 precision engineer's	20·00
	S400 and S440 precision engineer's with glass circle	23·75
	S771 automatic	18·25
	S770 automatic with glass circle	20·00
	S500 and S550 with built-in PP micrometer	23·50
Wild Heerbrugg	N0 builder's	13·00
	NK0 builder's with glass circle	14·75
	N2 engineer's	18·25
	NK2 engineer's with glass circle	23·75
	N2A automatic	18·25
	NK2A automatic with glass circle	23·75
	N3 precise with built-in PP micrometer	23·50
Zeiss Oberkochen	N1 4 small automatic	9·50
	N1 3 dumpy	14·75
	N1 2 precise automatic	18·25
	N1 2 precise automatic with glass circle	21·75

Table VI *Typical charges for hiring instruments*

Instrument	Hire service charge		
	per day	**per week**	**per month**
Levels			
Kern GK0 builder's level with tripod	£2·00	£3·50	£7·00
Kern GK1 engineer's level with tripod	£2·50	£5·00	£10·00
Kern GK1-A engineer's automatic level with tripod	£3·50	£7·00	£14·00
Kern GK23 engineer's precision level with tripod	£3·50	£7·00	£14·00
Geodetic levels with invar staves	£7·00	£15·00	£30·00
8in, 12in, 14in dumpy levels with tripod	£1·00	£2·50	£5·00
Theodolites			
Kern K1-A automatic theodolite	£9·00	£18·00	£35·00
Kern DKM1 theodolite	£9·00	£18·00	£35·00
Kern DKM2 theodolite	£14·00	£28·00	£55·00
Hilger & Watts microptic theodolite	£9·00	£18·00	£35·00
Vernier theodolite (all makes)	£4·00	£8·00	£15·00

Table VII *Typical rental charges for instruments*

Type of contract	**Rental service charges per £100 value per month**				
	1st year	**2nd year**	**3rd year**	**4th year**	**5th year**
Rental with insurance	£5·00	£5·00	£3·50	£3·00	£2·40
Rental with insurance and free maintenance	£5·60	£5·60	£4·60	£4·00	£3·50
Rental with insurance and free maintenance, plus loan of instrument during service	£6·00	£6·00	£5·50	£4·50	£4·50

Table VIII *Electronic Distance Measuring Instruments (EDMs)*

Manufacturer	**Type**	**Characteristics**
AGA	Geodimeter Model 6B	Measures distances within a known 200m range up to 5km in daylight, up to 25k, at night with a read off accuracy of ± 10mm. The Geodimeter transmits and receives a modulate light beam that is reflected from a distant target. Modulated light is insensitive to meteorological factors and electro magnetic disturbances. Time per reading 3min. Weight with case tripod, lamp and generator 46kg, reflectors extra
AGA	Geodimeter Model 8	For long rang work, in excess of 60km uses a gas laser to produce an emmited light beam of very low divergence, about 1m per 10km. Accuracy is ± (5mm + one millionth of the distance). Most of the error is due to meteorological factors. Time per reading 5 to 10min. Weight 43kg, reflectors extra
Zeiss	Electronic recording tacheometer, Reg Elta 14	This instrument measures distances up to 2km. It is designed for close range work which is achieved with economy and speed, 30 seconds for angle and distance measurement.

Manufacturer	Type	Characteristics
		Weight of instrument case tripod and accessories 47kg, reflectors extra
Wild	Distomat D1 10	Transmits a harmless infra red carrier wave and can measure distances up to 2km with an accuracy of ±10mm. Measuring time is 15sec. The D1 10 may be mounted on a suitable theodolite to form a distance and angle measuring unit. Weight, excluding theodolite and tripod, is 22kg, reflectors extra
National Physical Laboratories	Mekometer	Still undergoing development this instrument will be made under licence from the NPL. It promises to be a keen rival of established EDM's
Tellurometer Ltd	Tellurometer MRA 101	Tellurometer Ltd specialise in EDM's that transmit and receive a microwave signal. The MRA 101 has a range of 15km with an accuracy of ±1·6cm per km

Part 2
Surveying techniques

Surveys of land as they concern architects are usually done to establish boundaries of sites, or to locate accurately natural features such as trees, ponds, streams, steep banks, and so on, or artificial features such as buildings, embankments, roads, pylons etc.

The common and essential characteristic of surveying techniques, other than air photography, is that a series of lines is laid out on the ground in a geometrical shape which can be reconstructed to scale on a drawing board. The purpose of such geometrical shapes is to set up a reference framework that can be accurately reproduced on a drawing and from which irregular shapes can be easily identified by a series of minor measurements (known as offsets). These are usually made at right angles to the main lines of a survey at measured distances along them.

Surveying techniques

Two techniques exist for establishing a geometrical outline or perimeter for a survey framework as described above:

1 the laying out of a series of interrelated triangles in which the lengths of the sides are known: a technique known as *triangulation*;

2 the laying out of a series of lines of known length and angular relationship: a technique known as *traversing*. The lines of a traverse need not outline a geometrical figure as with triangulation, but it is desirable, whenever possible, that they should do so.

Triangulation is the simplest method geometrically but may become very laborious in practice if many triangle sides serve no other purpose than for setting out. On the other hand in its most elementary form it requires no instruments for measuring angles.

In wooded or hilly country, in towns, or wherever there are many obstructions it may become impossible to lay out a series of interrelated triangles. In such cases, or where long boundaries enclose a more or less blank area, it is best to resort to traversing.

In some cases it will be found that traversing is the best technique for part of a site but that triangulation can be used for the rest. Choice of technique depends upon the circumstances. The following is a general guide.

Surveying by triangulation

1 Open sites of up to one hectare.

2 Open or fairly open sites with detail that needs recording well within the boundary.

3 Checking that rectilinear sites are rectilinear, and if not what shape they are.

4 Tying in detail (such as buildings) to a known line, an existing straight boundary, a building site, a straight road, railway or canal.

5 Establishing accurately the location of a number of points over a large area to which surveys of smaller areas can be related.

Surveying by traversing

1 In towns and cities, or wherever it is possible to see along only two intersecting lines at a time.

2 In country where irregularities of land such as hills obscure cross sights to opposite corners of an area.

3 Where long boundaries enclose more or less featureless land.

Chapter 9
Surveying techniques: Triangulation

This chapter gives an account of the principles of triangulation as a technique for carrying out topographical surveys. It is intended to be read in conjunction with Chapters 4 *to* 7

Triangulation

One of the simplest and most satisfactory methods of topographical survey is to mark out triangles over.

Triangulation has the advantage that it can be done over small areas with simple instruments and without measuring angles. For this reason it is probably the commonest method employed for small land surveys. It is also used, wherever possible, for larger surveys with angular measurements because it cuts down the possibility of error.

The need for angular measurement also obtains in difficult areas where it might be impossible to make physical measurement with chains or tapes or over irregular land where chaining might be inaccurate on account of excessive slopes.

All the methods of linear measurement previously described can be used for surveying by triangulation but the choice must be governed by the degree of accuracy required. Some possible combinations of method listed below in increasing order of accuracy are intended as a guide.

Methods of carrying out triangulation surveys

1 Plane table with telescopic alidade using tacheometry. (Plane tables and telescopic alidades are described in Chapter 2.6. Tacheometers are described in Chapter 2.2. Tacheometry is dealt with in Chapter 5.

2 Chaining with metric or engineer's chain. Metric and engineer's chains are described in Chapter 2.2. Procedure for chaining is dealt with in Chapter 5.

3 Plane tables as above plus steel tape for main dimensions. (Steel tapes are described in Chapter 2.2 and their use in Chapter 5.)

4 Chaining as above plus horizontal angles with a level or small theodolite. (Levels are described in Chapter 2.5, theodolites in Chapter 2.7. Their use in measuring horizontal angles is described in Chapter 6.

5 Chaining with steel band or steel tape. (See Chapter 5.)

6 Horizontal angles with a theodolite plus accurate baseline measurement with a steel band or tape. (See Chapters 4 and 6.)

Triangulation networks

The simplest possible case of a triangulation survey is that in which the area concerned is roughly triangular **1**. Depending upon size this could be surveyed by three main chain lines and offsets, or three main chain lines and subsidiary triangles outside the framework of the main triangle.

The most usual case encountered is a roughly rectangular piece of land whether it be the whole or part of a survey. In such instances some combination of two triangles is normally possible. **2** illustrates the general idea with two triangles ABD and BCD.

This figure raises an important point: It is not sufficient to regard a diagonal like BD as a guarantee of accuracy. If BD is measured incorrectly, especially by a comparatively small amount, the whole figure ABCD will be thrown out of shape when the survey is plotted but it will not usually be apparent. Check lines such as AC in **2** should be measured whenever possible, in addition to those absolutely necessary for plotting the survey. Some other possible ways of providing check lines for ABCD are also shown in **3** and **4**.

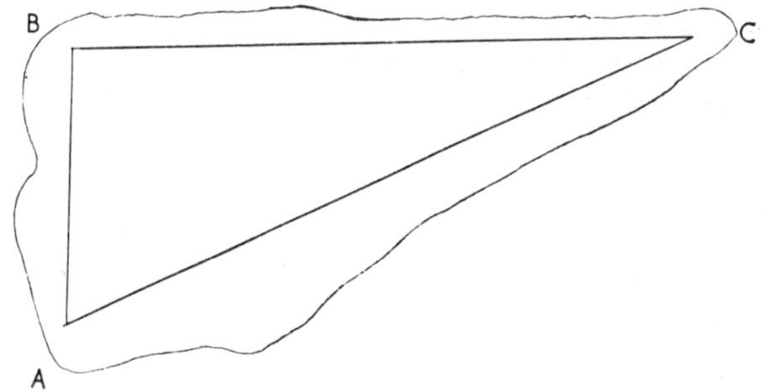

1 *Triangle of chain lines laid out oc a roughly triangular plot so that the boundaries can be fixed by short offset measurements made at right angles along each chain line at suitable intervals*

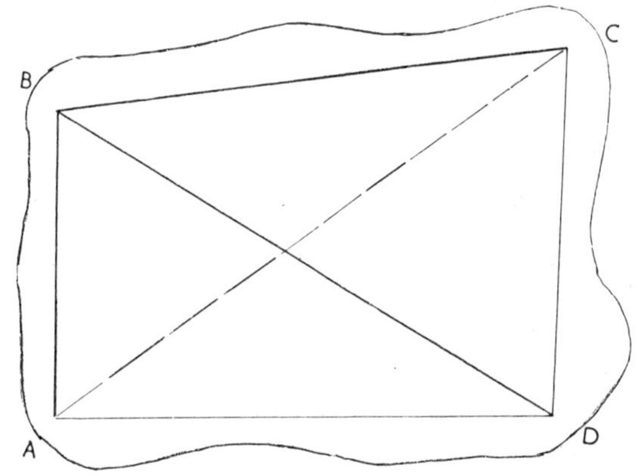

2 *Two triangles of chain lines laid out to plot a roughly rectangular piece of land. Short offset measurements are made from lines AB, BC, CD, DA to establish the boundaries. AC should be measured as well as BD to form a complete check on shape*

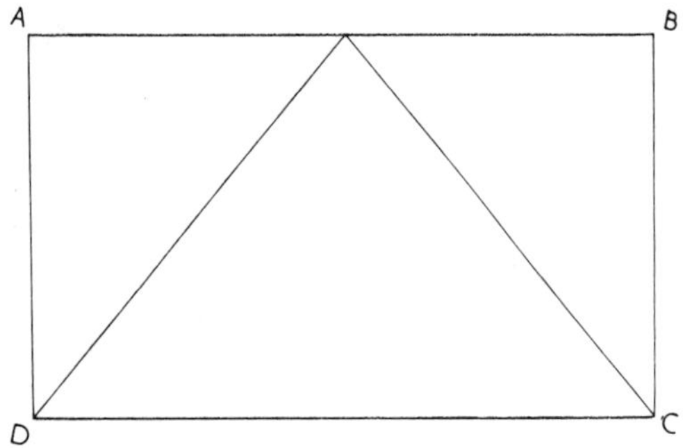

3 *Alternative system of triangles for a roughly rectangular piece of land*

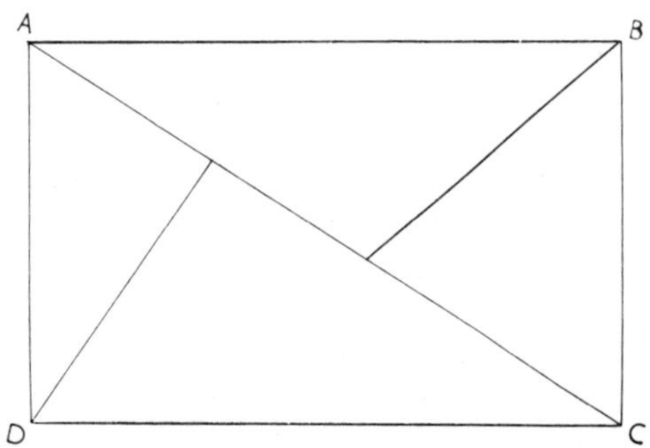

4 *Alternative method of checking shape for two triangles on a roughly rectangular piece of land. The lines from B and D respectively to the diagonal are the check lines*

Procedure in the field

Procedure for triangulation should be along the following lines: reconnaissance; setting up station points; chaining or making instrument observations.

Reconnaissance The surveyor should walk over the area to be surveyed noting any points of special interest and making a rough sketch of the area or checking it against an existing drawing or map.

Station points are set up by marking the terminations of chain lines with ranging rods, pegs, or surveyor's nails. The choice of points should be made with the following considerations in mind:

There should be the fewest possible station points.

At least one line should run right across the middle of the area.

Triangles should be as nearly equilateral as is practicable ie 'well-conditioned'.

As many lines as possible should be planned to obtain full value from offsets.

Offsets should be short.

Lines should be chosen to avoid obstacles as much as possible.

Lines should be as level as possible.

Lines should be visible from end to end if possible.

In practice various kinds of obstacles are encountered which, if overcome, obviate the need for extra triangulation. Obstacles fall into three categories:

1 Those which obstruct vision but not linear measurement.

2 Those which obstruct linear measurement but not visibility.

3 Those which obstruct both measurement and visibility.

Obstacles in category 1 are usually humps high enough or hollows deep enough to prevent the rear chain man from directing the front chain man on to the terminal point of a chain line **5**.

Procedure for overcoming both conditions of category 1 as illustrated in **5** is shown in **6**.

Obstacles in category 2 are, typically, areas of water exceeding the length of a chain or tape. The problem arises in two forms.

(a) Where it is possible to chain around the obstacle.

(b) Where it is not possible to chain around the obstacle.

Two possible methods of dealing with (a) are illustrated in **7** and **8**.

There are other methods for dealing with these problems but these have been chosen as typical and useful since they facilitate offsets.

One possible method for dealing with obstacles

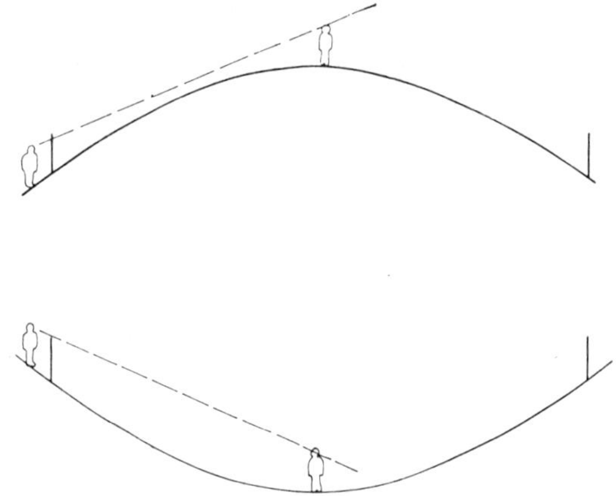

5 *Two examples of chairing in which natural obstacles (humps or hollows) prevent marking out a chain line with poles*

which cannot be chained around is illustrated in **9**.

Obstacles in category 2 do not affect tacheometrical measurement.

Obstacles in category 3 may be buildings or thickets or other similar objects. The solution (even in tacheometric surveying) generally requires fairly complex direct linear measurement to prolong the chain line beyond the obstacle. Two useful solutions among a number that are to be found in textbooks are illustrated in **10** and **11**.

It should be noted that neither of these solutions produces particularly reliable results and obstacles

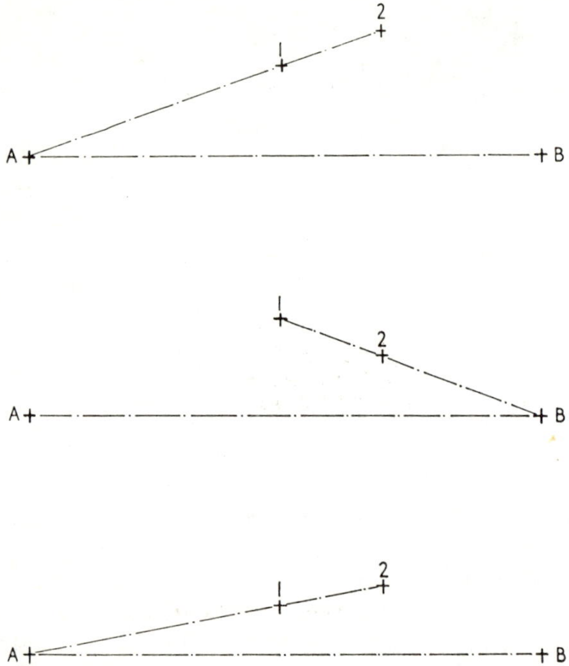

6 *Procedure for overcoming conditions shown in 5: The two chain men make sure the ranging rods at A and B on plan are truly plums and then pick two intermediate points, 1 and 2, as nearly as they can judge on the chain line AB. No 2 sights No 1 into A; No 1 sights No 2 onto B; No 2 again sights No 1 onto A. This is continued until both No 1 and No 2 are satisfied that their opposite number's ranging rod is in line. Measurement can then be made by chaining or tacheometry*

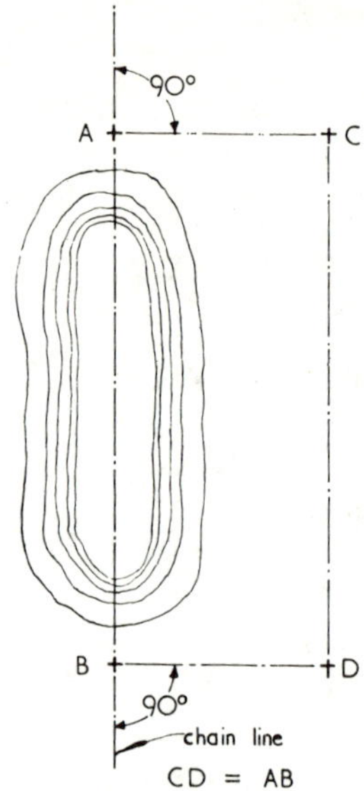

7 *The chain line AB can be side-stepped by means of right angled offsets AC and BD of equal length. CD then equals AB and can be measured*

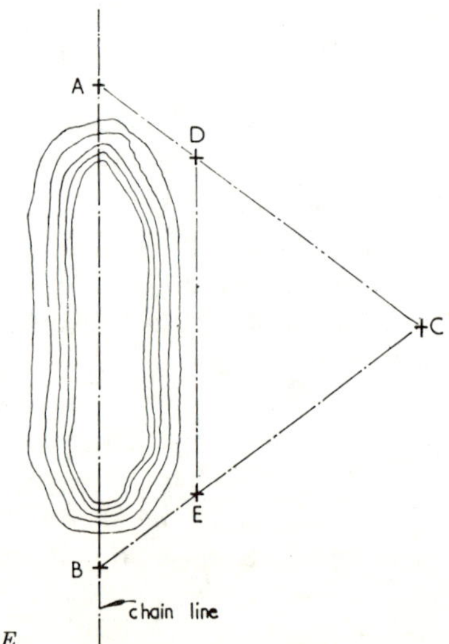

8 *To determine length of AB select a point C and measure AC and BC*
Choose a point D on AC and find the value $n = \frac{ca}{cd}$
Find a point E on CB such that $CE = \frac{cb}{n}$
Measure DE then $AB = n.\ DE$

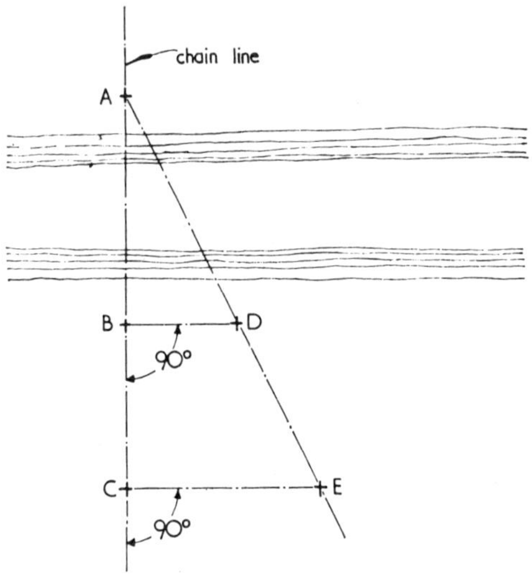

9 *To determine length AB procedure is as follows:*
From two points B and C on the main chain line set out perpendiculars
Select a point D on the perpendicular from B.
Find a point E on the perpendicular from C such that ED and A are in line.

Measure BC, BD and CE.

Then $AB = \dfrac{BC \times BD}{CE - BD}$

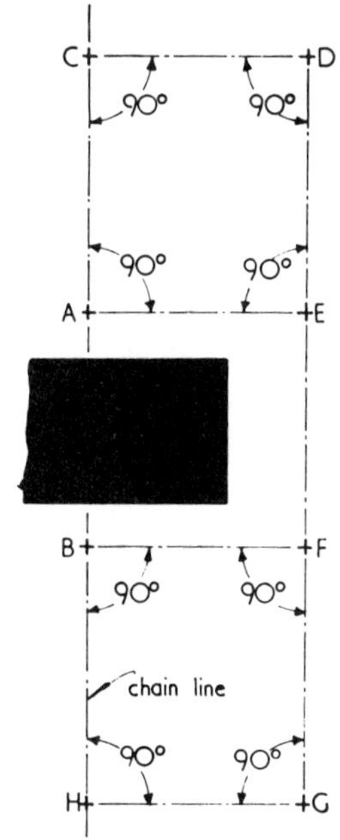

10 *To determine AB first set up two equal perpendiculars at C and A, thus fixing two points D and E parallel to the chain line. Two further points, F and G, are now sighted in to extend this line DE to G. At F and G two perpendiculars FB and GH are next set up each being equal to CD and AE. Thus EF = AB, the required dimension and BH is in line with CA*

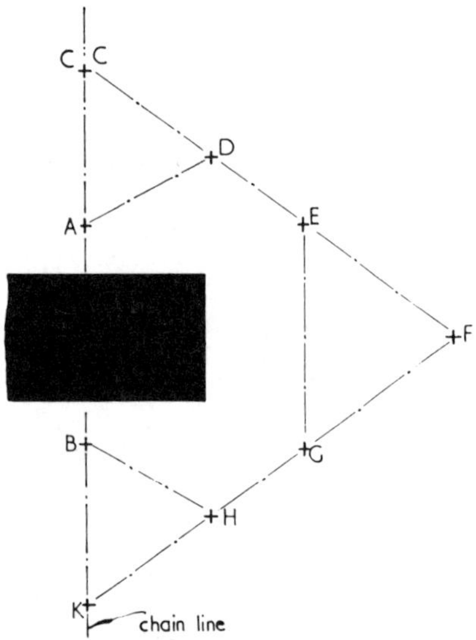

11 *To determine AB first set up an equilateral triangle ACD. extend CD to a convenient point F. Set up an equilateral triangle EFG where E is a point on the line CF. Extend FG to K so that FK = CF. Set up an Equilateral triangle KHB where H is a point on the line FK. Then AB = CF − AC − BK and an extension of BK will be in line with CA*

in category 3 should be avoided unless an instrument for angular measurement is available.

Field records

The records of a triangulation survey should include a sketch map of the area covered with all chain lines shown **12** including the direction in which they were measured. Further information which should be noted either in the sketch plan or chain line notes, is given in the check list below.

Check list of information to be noted on survey drawings

North point. (See also Chapters 10 and 11). Nature of ground: lawn, pasture, arable, scrub and so on.
Direction of flow and depth of rivers or streams.
Depth of wells or pits.
Species of trees if large enough for sale as timber.
Type and construction (in general terms) of any buildings.
Signposts and milestones.
Names and numbers of all houses or other properties.
Names of streets and Ministry of Transport numbers for classified roads.
Direction and destination of roads, footpaths and railways.
Hydrants, manholes, post office cable markers.

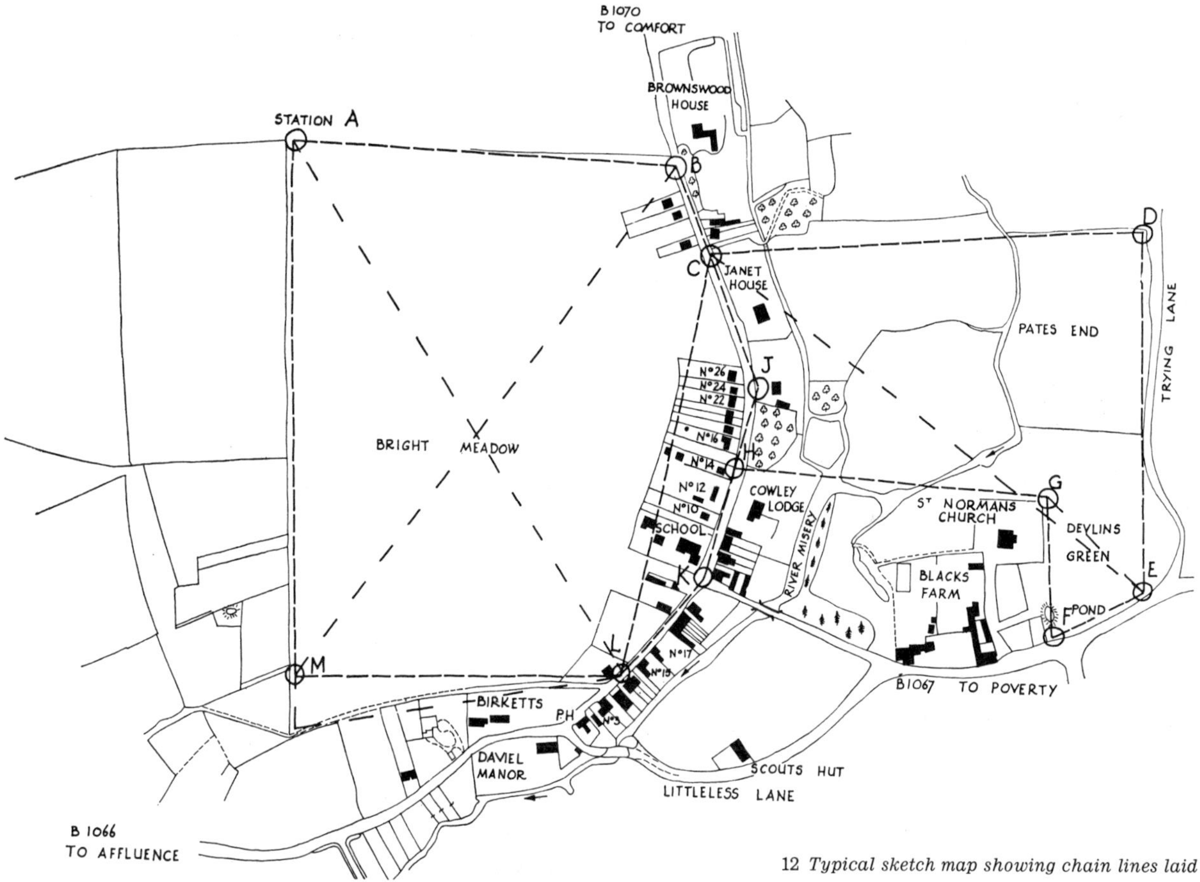

12 *Typical sketch map showing chain lines laid out*

Lamp posts, telegraph, telephone and electricity supply poles.
Boundaries and their nature; hedge, fence, ditch, wall and so on. Gates, width and direction of opening.
Any other special features.

Field books

Apart from the sketch map, which serves only as a key, the compilation of field records is done in field books. These may be purchased and are usually in the form of a book about 220mm × 120mm opening lengthwise.

The pages are ruled centrally lengthwise with either a double or single line. This latter represents the chain line into, or across which, linear dimensions are written as they are measured.

The space either side of the central ruling is used for sketches of features located from the chain line (eg offsets) notes and intersections with other lines. Subsequent diagrams illustrate typical examples of the method.

It is important to adopt a standard procedure for booking in any one practice because it may be necessary to refer back to records long after everyone has forgotten what method was used to indicate what sort of measurement and so on was booked.

The field book is started at the back and measurements are recorded progressively up the pages. The first point on a chain line should be marked with a zero and circled **13**. Similarly any station points recurring on the chain line should be listed and circled.

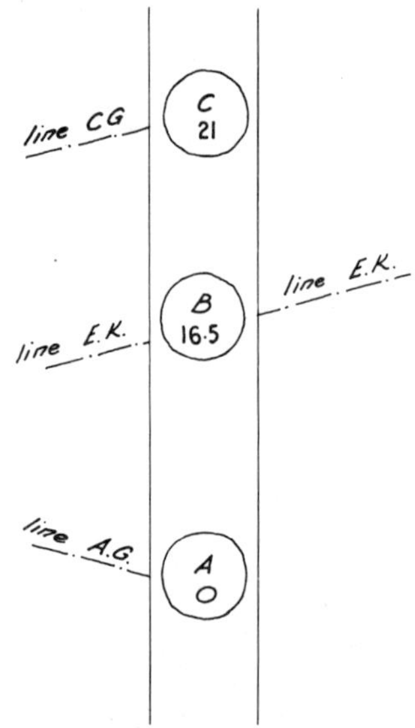

13 *Station point recorded in a field book by circling the station reference and dimension, starting from zero at the bottom of the page Intersecting chain lines should be clearly labelled*

Offsets are recorded by at least two dimensions: distances along the chain line and distance from the chain line. The one should be opposite the other **14**.

A means of recording comparatively large features such as houses and gardens which cannot be fixed satisfactorily by offsets is shown in **16**.

Trees of any size or importance should be located to the approximate centre of their trunks. Their spread should also be indicated and roughly dimensioned (an optical square can be used to locate rough zenith lines) together with an estimate of height and species **18**.

A typical sheet from a field book is illustrated in **17**. It should be noted that at points of intersection with other lines the latter are indicated as meeting and leaving the double ruled 'chain line' in line with the number that indicates position along the chain line.

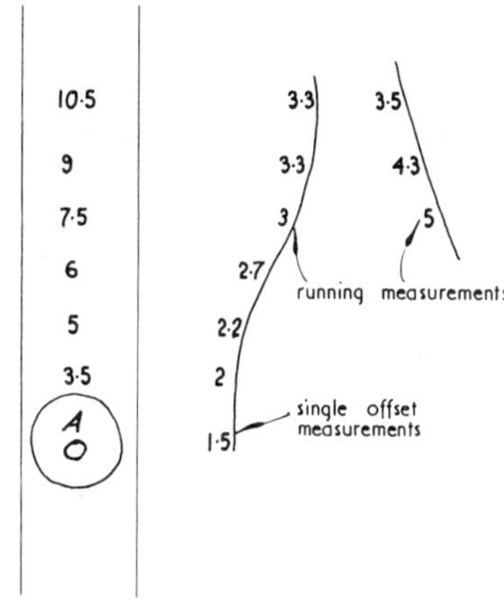

14 *Offset measurements from a chain line shown by dimensions of perpendiculars placed opposite the figures representing distance along the chain line. Running measurements along offsets indicated by writing figures parallel to the chain line*

Triangulation with horizontal angles

It becomes much easier to check triangulation if angles can be taken at station points. This can obviate the need to take linear measurements except where offsets are required, since it becomes possible to calculate values trigonometrically.

19 an example of a field book in which angles have been recorded. Normal practice is to put these on the right-hand side but the example has been condensed for illustration.

The angles shown in **19** represent four readings for each angle. Two each on a pair of verniers at 180deg and one set 'face left' and one 'face right' (ie with the vertical circle of the theodolite on the left of the telescope and then with the telescope transitted, putting the vertical circle on the right). The final values for each angle are the means of the four readings.

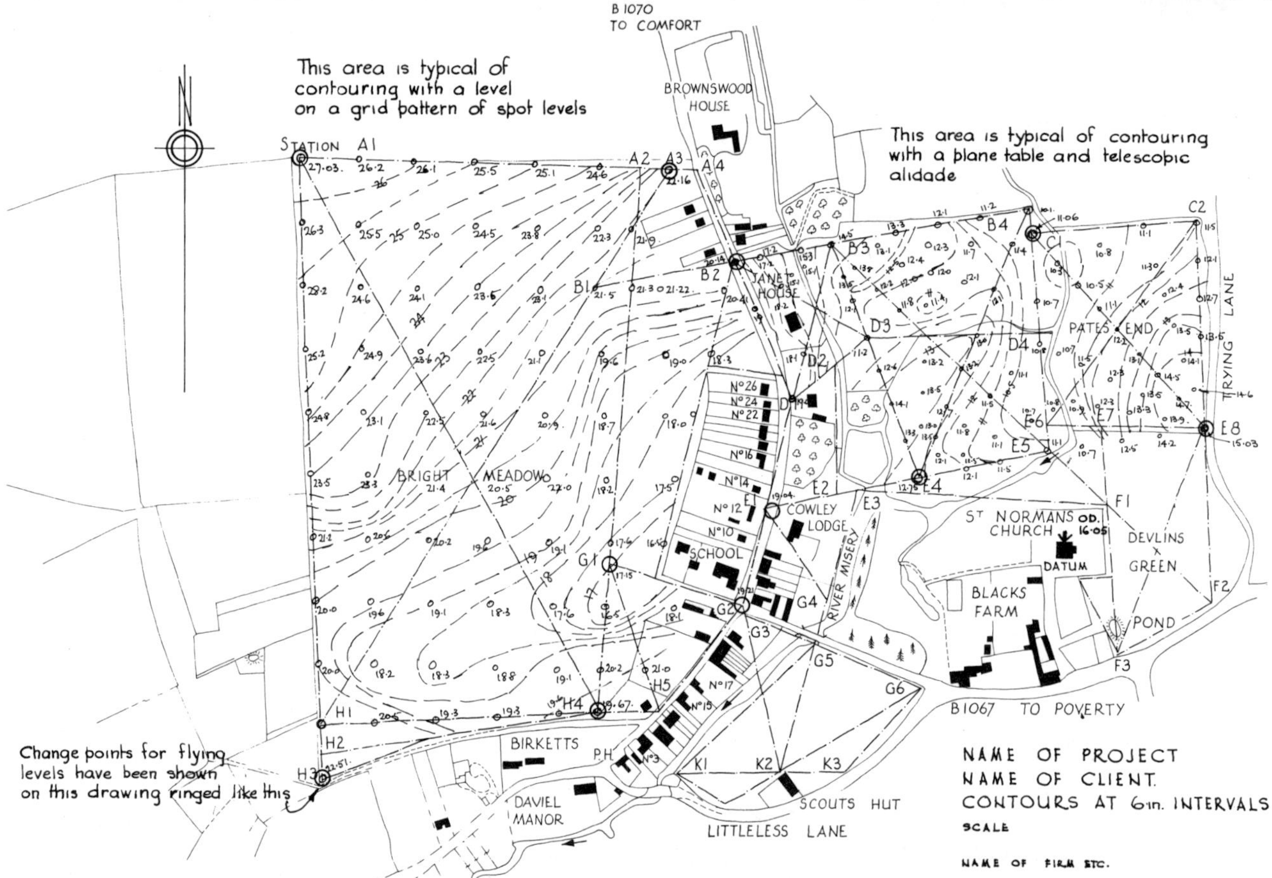

15 This diagram represents the same survey map as 12 rearranged for a traverse survey (see Chapter 11) The arrangement for triangulation using angular measurement would be very similar. Field records and field books are the same for triangulation and traversing. The diagonal lines are angular sightings without linear measurement

Triangles can be checked by adding all the angles (total 180deg), and calculated **20** using the formula:

$$\frac{a}{\sin A} = \frac{b}{\sin B} - \frac{c}{\sin C}$$

In practice there are more convenient methods of calculation, some of these are given in Chapter 13.

Boundaries

Measurements made to boundaries when chaining or traversing must represent the limits of ownership when two properties adjoin. Some typical examples of demarcation are shown in **21**.

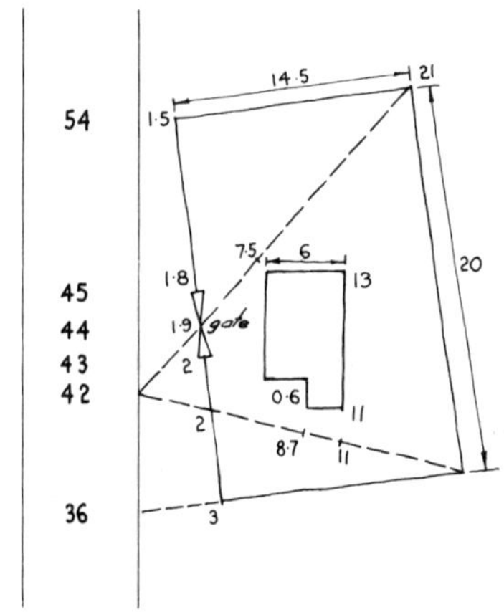

16 *Larger items such as houses, gardens and so on, can be fixed by subsidiary survey lines as well as offsets. Dimensions on subsidiary lines are written right angles to them when recorded as running dimensions*

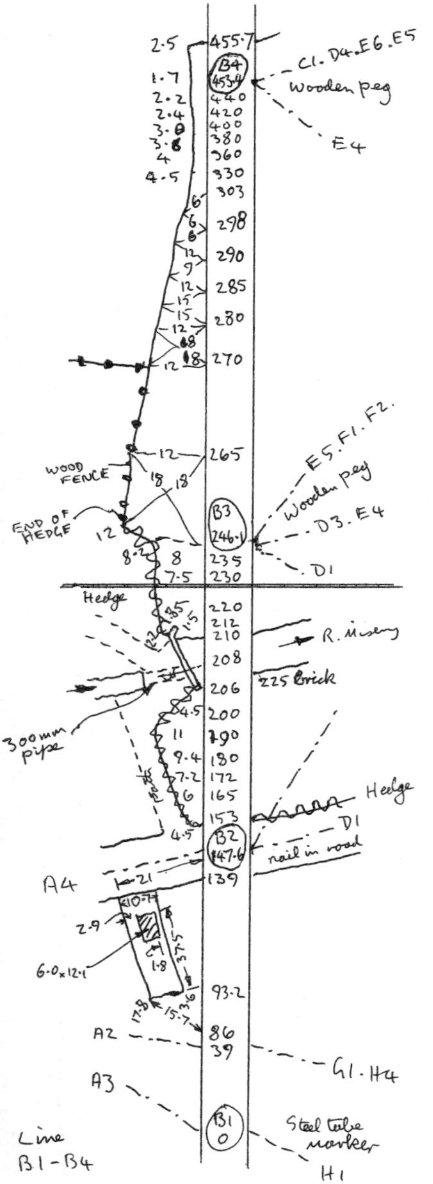

17 *Typical page from a field book. This is an example of the field record of chain line B1 to B4 shown in 12*

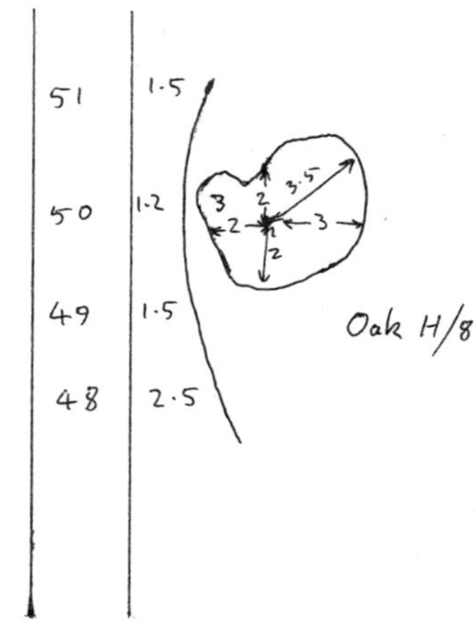

18 *Details of large trees should be recorded with approximate dimensions of spread from the trunk indicated*

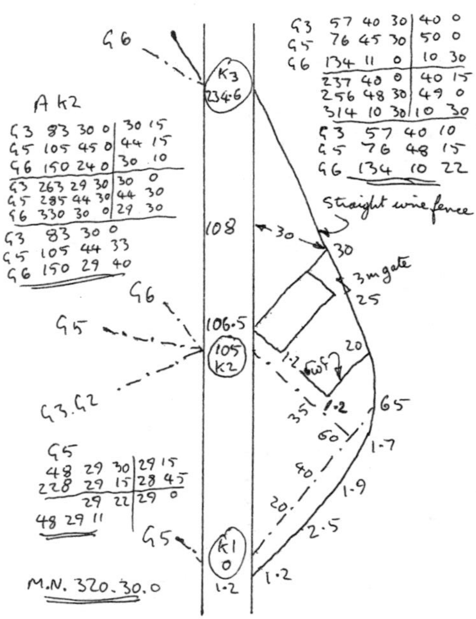

19 *Example of a page from a field book in which angles have been recorded. This data refers to chain line K1 to K3 in 12*

$$\frac{K1:K3}{\sin K1.G5.K3} = \frac{G5:K1}{\sin G5.K3.K1} = \frac{234{\cdot}6}{\sin 54.42.34} = \frac{G5:K1}{\sin 76.48.15}$$

log 234·6	2·3703
log sin 76.48	$\bar{1}$·9884
	2·3587
log sin 54.42	$\bar{1}$·9118
log G5:K1 =	2·4469
G5:K1 =	279·9

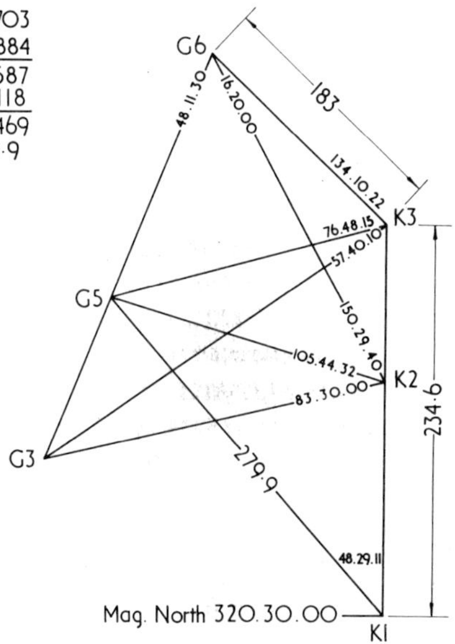

20 *Part of the triangulation illustrated in 12 and the calculation for determining the dimensions of area of the triangules contained in it*

J

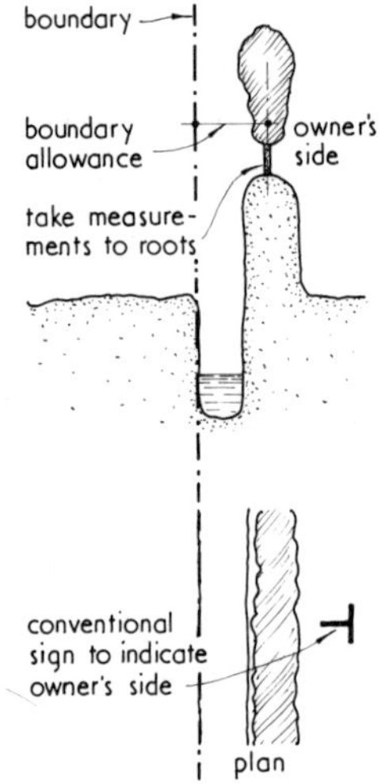

ditch and hedge (or fence)

note: it is customary to consider an owner's boundary extends to the brow of the ditch in a case of this kind

the boundary allowance varies in different parts of the country and in various circumstances; enquiry into local custom should be made

average boundary allowance				
land		metres	feet	links
belongs to	same owners	1	3	5
	different owners	1·2	4	6-7
adjoins public land		1·2–2	4-6	7-10

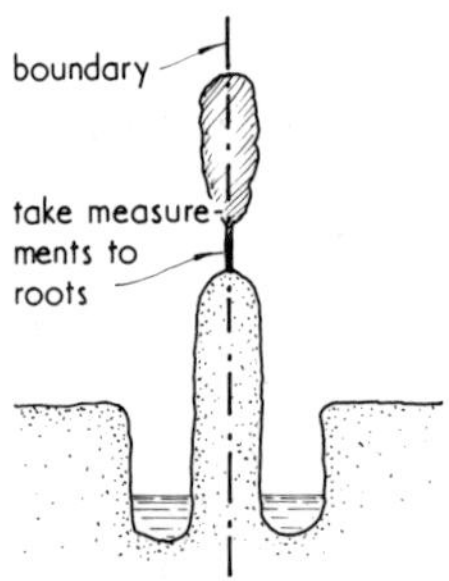

two ditches and hedge (or fence)

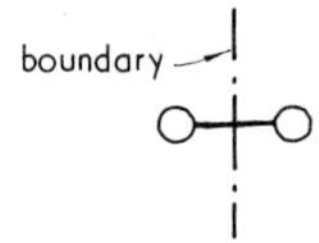

conventional sign to show change in the formation of the boundary

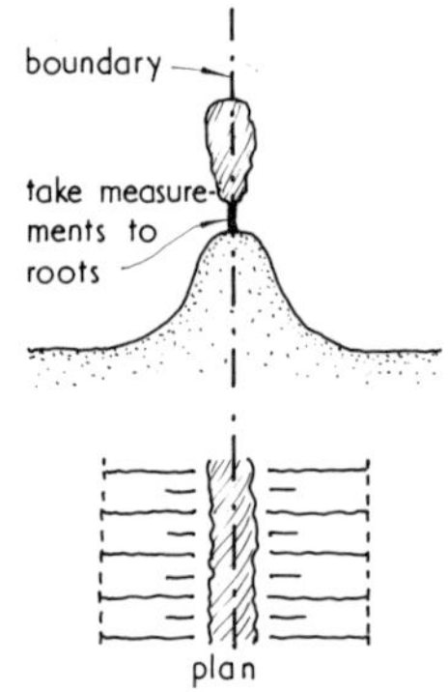

bank and hedge (or fence)

note: with a fence, the face of it constitutes the real boundary; the owner looks upon the back of it and nails are driven (towards) 'home'

21 *Typical cases of property boundaries to indicate the points to which offset measurement should be made*

Chapter 10

Surveying techniques: Bearings and coordinates

This chapter provides an introduction to the Ordnance Survey National Grid system of map reference showing how it assists in carrying out surveys and plotting results. It also provides an elementary explanation of the principles of plotting surveys by calculated co-ordinates from angular bearings and linear measurement. For a description of the services provided by Ordnance Survey see Chapter 15

General

It would be almost impossible to overstate the importance of the concept of location by bearings in the development of the modern world. It transformed navigation and provided the basis for the most outstanding work of survey ever carried out without the help of aeroplanes: the first Ordnance Survey of Great Britain begun over a hundred years ago. It is still one of the most useful theoretical tools available for survey, even over small areas.

The theory of bearings and co-ordinates can be applied to both triangulation and traversing. It is particularly useful when plotting surveys and setting out buildings. (See Chapters 9, 11, 13 and 14.)

For the purposes of navigation, map-making and geodetic survey the earth is considered to be covered by a reference network of imaginary lines of latitude and longitude.

Lines of latitude are a series of rings concentric with the earth's north-south axis. Lines, or *meridians*, of longitude are great circles drawn encircling the earth, passing through the north and south poles and cutting the equator at right angles to it. The equator is considered to be divided into 360 intervals (ie degrees) by the meridians that cut it. The meridian passing through Greenwich is generally taken as the original reference and designated 0 (zero) degrees.

Meridians of longitude converge towards the poles; thus at latitudes north of the equator the linear distance between meridians becomes less as one proceeds north, but the angular difference remains the same, being reckoned from the earth's north-south axis.

Various methods or *projections* have been used to try and overcome the difficulty of representing the spherical surface of the earth on a flat plane of paper. Meridians are, of course, curved when thus represented over large areas but may be considered as virtually straight over small areas. This principle has been used by Ordnance Survey to establish the National Grid system—which is the one most important for architects and surveyors in Britain to know.

The National Grid

The National Grid system comprises a grid or network of 100km squares covering the whole of Britain and lettered (originally numbered) for reference. On 1in to 1 mile maps the grid is further subdivided down to 1000 squares on which points can be located easily to within about 100m. These maps provide the index to the 1: 25 000 series of Ordnance maps and will in the future perform the same function for larger scale maps.

Detailed information is available from the Ordnance Survey, Romsey Road, Maybush, Southampton S09 4DH.

Meridians

Bearings or angular orientation of one object relative to another can be established by using one of the following reference directions:

1 true meridian

2 magnetic meridian

3 grid meridian or

4 an arbitrary meridian.

True meridian is the meridian of longitude. It is usually determined by astronomical observation and as such is beyond the scope of this book. However it can be found, with latitude, to within about 6sec and without much difficulty from 1in Ordnance Survey maps, should it be needed. True meridian at any point is invariable and thus provides a permanent reference.

Magnetic meridian is magnetic north as determined by a compass. It varies continuously and differs from place to place and thus lacks the advantage of a permanent reference. One-inch Ordnance Survey maps usually provide data which enable the true meridian to be estimated from compass observations.

Grid meridian, in Britain, is derived from a reference true meridian (2deg west) by drawing plane parallel lines to form the north-south lines of the National Grid. Thus, at any point not coincident with true meridian, grid meridian is at an inclination to true meridian. Again, Ordnance Survey maps usually provide data so that grid meridian can be related to true meridian if necessary.

Arbitrary meridian. It is often convenient on a small survey to relate all bearings to some prominent permanent point observed from one survey station.

This has the advantage that it is easily rediscovered at a subsequent date for further operations. It is nevertheless advisable to record a magnetic bearing. In such cases it is usually quite easy to convert the arbitrary meridian to grid or true north should that become necessary.
It can be shown that the relationship between any three points can be established by the distance between two of them and the angle subtended by two of three at the third. It is also possible to establish the relationship between two points and all others within a reference grid by means of distance and a bearing from grid meridian **1**.

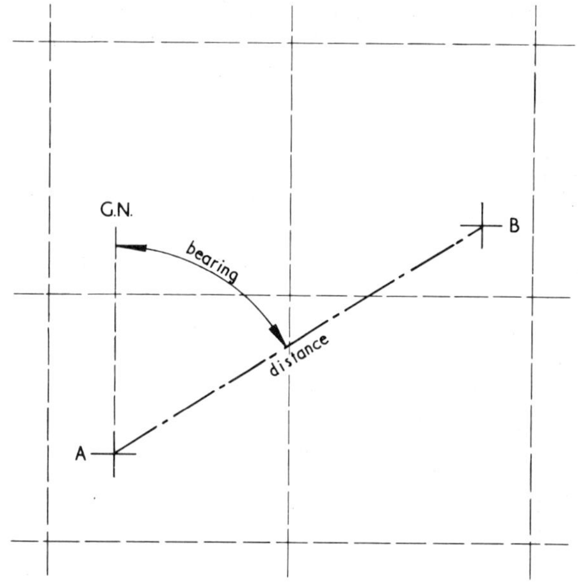

1 *The relationship of point B to point A is fixed by the bearing (angle) and distance AB. In the illustration A to GN (grid north) is parallel to the grid shown, in which the vertical lines represent north*

Bearings

There are two commonly accepted ways of designating bearings when recording angles such as GN A B in **1**. These are:

1 the whole circle system,
2 the quadrantal system.

The whole circle system

The whole circle system of designating bearings takes north as zero (ie 360deg) and expresses all bearings as angles clockwise. Thus if the bearing A→B is x deg in **1** then the complementary bearing B→A would be 180deg + x deg.

The quadrantal system

The quadrantal system divides the whole circle into four quandrants of 90deg, in each of which angles are reckoned from either north or south. The system may be expressed as follows:

North

Angles = N x degrees W	Angles = N x degrees *E*
Angles = S x degrees W	Angles = S x degrees E

In **1** the bearing of A to B on the quadrantal system would be N x deg E and that from B to A would be S x deg W.

Both the whole circle and quadrantal systems offer their own advantages. Those of the whole circle system conform more closely to the graduations of most instruments likely to be encountered in practice and so the whole circle system has been adopted as standard for this book.

Back bearings

The complementary bearing referred to above is properly called a *back bearing*. To obtain the back bearing of any survey line from the second station ie, back to the station from which the *forward bearing* was observed (A in **1**), add 180deg if the forward bearing is less than 180deg or subtract 180deg if it is more.

When making calculations with books of tables in which trigonometrical functions are given only for angles up to 90deg it is necessary to reduce whole circle angles to trigonometrical equivalents in that range. These are called *reduced bearings*. They are explained in Chapter 11 where the method of recording angles is described.

Co-ordinates

There are various methods of plotting surveys which are described in the textbooks. Some of these methods are given in Chapter 13, but the system which most conveniently fits in with architectural practice—especially where some form of modular planning is concerned—is the co-ordinate system. It is also the most accurate.

The co-ordinate system of plotting surveys

The co-ordinate system employs a form of grid reference analogous to the National Grid. The area surveyed and plotted on drawings may be imagined as covered with a grid of lines a convenient distance apart, say 5m or 50m according to the size of the survey, and running parallel with the meridian (whether true, magnetic, grid or arbitrary). Any point in the grid can then be identified by two numbers representing distances east of the reference meridian and north (or south) of a reference latitude. The principle is shown in **2**.

The grid is most conveniently assumed to begin with a meridian *west* of any point in the area concerned and with a latitude *south* of it. The point at which meridian and latitude meet is called the *origin* and conventionally labelled 0 (zero).

Following Ordnance Survey practice the two dimensions are termed *eastings* and *northings* and by convention are given in that order so that it is, in fact, sufficient to give the two numbers separated by a point or preferably a stroke. Thus the co-ordinates of station A in **2** could be written 75/120. Surveyors commonly use the terms *latitude* and *departures* in place of northings and eastings respectively, while Continental surveyors often employ mathematical notation, calling the reference latitude the x axis, the reference meridian the y axis and distances from them the x and y co-ordinates.

When plotting a survey it is usually convenient to locate the reference meridian and reference latitude so that they lie just outside the western and southern limits of the survey. This has the effect of placing the origin near the bottom left-hand corner of the survey drawing. Its precise location can be chosen by assuming co-ordinates for the most south-westerly point of the survey.

The direction of the grid reference meridian can be fixed equally easily either (*a*) by finding magnetic north at the station point for which co-ordinates have been assumed (*b*) by making it parallel to a chain line from the same station point, or (*c*) by sighting on a convenient permanent and prominent object such as a church weather cock.

Calculations for locating one point relative to another in a co-ordinate system, given the bearing and distance between them is very easily done by trigonometry. They consist mainly in the solution of a series of right angled triangles of the sort shown in **3**. Methods of calculation and graphical plotting are discussed in Chapters 13 and 14.

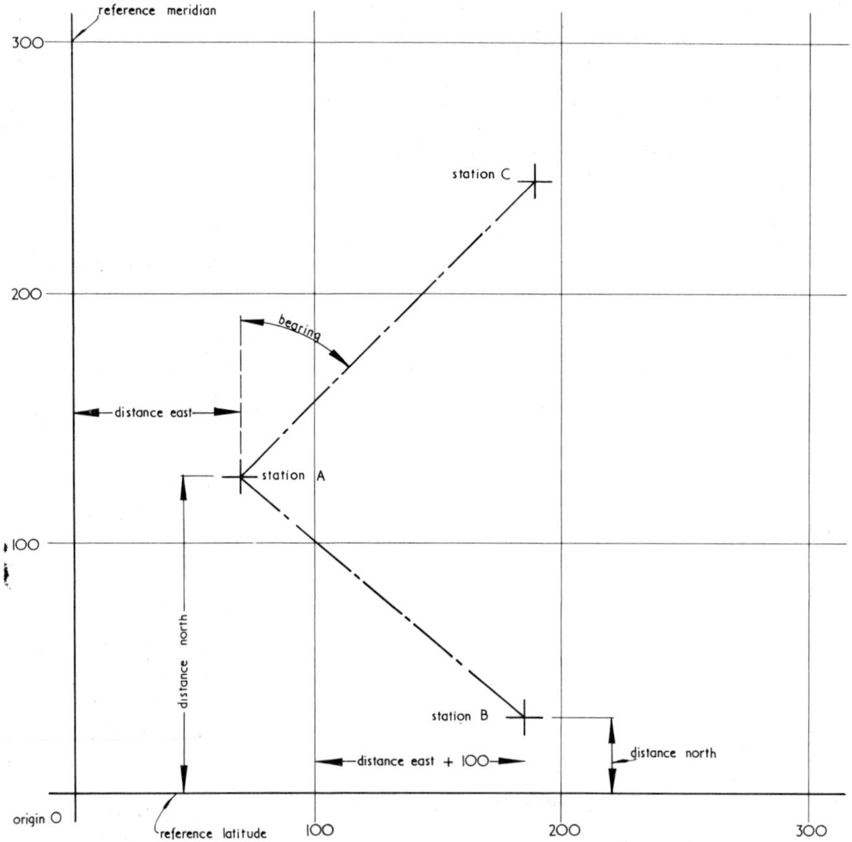

2 The location of any point in a co-ordinate grid can be fixed by means of a distance east and a distance north. Thus stations A, B and C might be identified as follows: A = 75 east 120 north; B = 180 east 40 north; C = 190 east 250 north

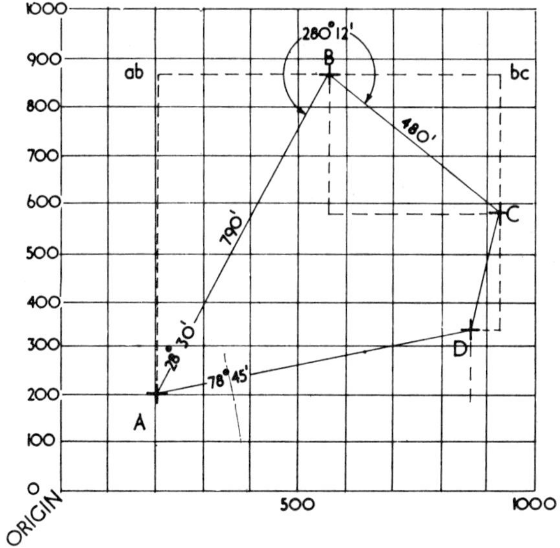

Coordinates of A : 200 E, 200 N
Angle ab.A.B = 28° 30' Bearing = 28° 30'
Distance A B = 790 m

Coordinates of B by calculation — East ab.B = 790 sin 28° 30'; North Aab = 790 cos 28° 30'

log 790	2·8976	log 790	2·8976
log sin 28° 30'	$\bar{1}$·6789	log cos 28° 30'	$\bar{1}$·9439
log ab.B	2·5763	log A.ab	2·8415
ab.B = 377·0 east		A.ab = 694·2 north	

Coordinates of B : 577 E, 894·2 N
Angle A.B.C = 280° 12' Back-bearing B.A = 208° 30'
Distance B.C = 480 m Bearing B.C = 128° 42'

Coordinates of C by calculation — East B.bc = 480 cos 128° 42'; North bc.C = 480 sin 128° 42'

log 480	2·6812	log 480	2·6812
log cos 128° 42'	$\bar{1}$·8923	log sin 128° 42'	$\bar{1}$·7960
log B.bc	2·5735	log bc.C	2·4772
B.bc = 374·5 east		bc.C = 300 north	

Coordinates of C : East 577 + 374·5 = 951·5
North 894·2 − 300 = 594·2

3 *An example of the method of calculation for locating one point relative to another in a co-ordinate system*

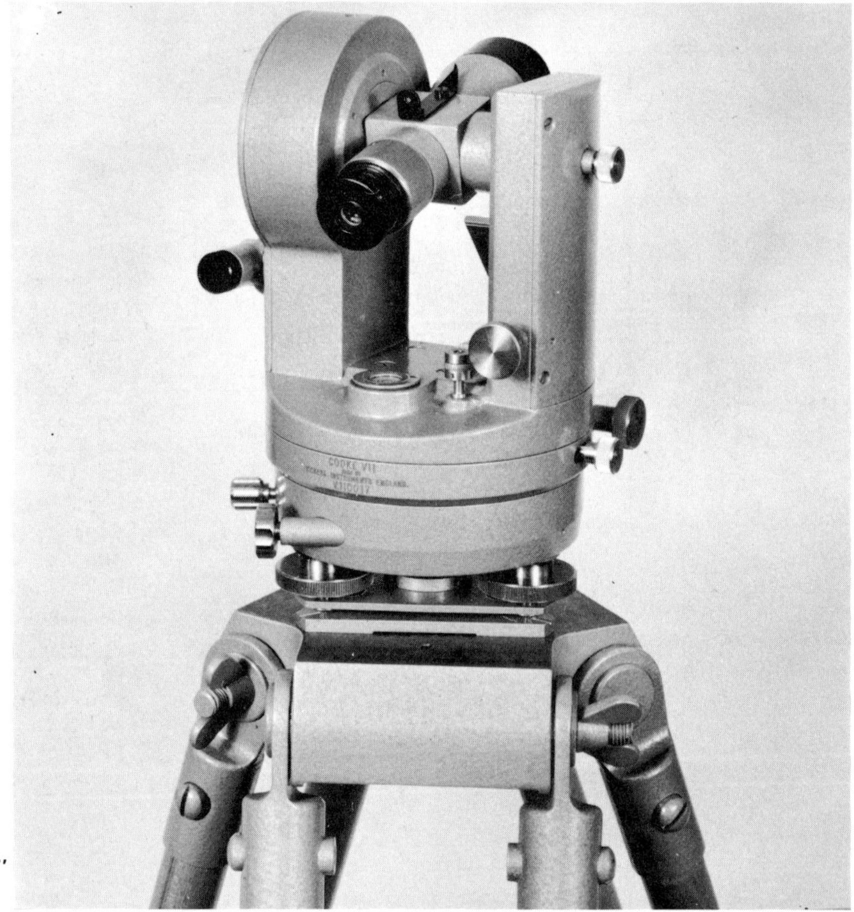

Chapter 11

Surveying techniques: Traversing

This chapter describes the technique of traversing in carrying out a topographical survey. It is a technique involving linear measurement, the measurement of angles and the plotting of bearings and co-ordinates. It can be used instead of or in conjunction with the technique of triangulation

Traversing

Traversing is a method of fixing the location of a series of station points by means of distance and bearing measurements. The technique thus depends upon linear measurement and measurement of horizontal angles.

When making traverse surveys the surveyor starts at one point, measures the distance to a second, and fixes the direction from that to a third by measuring the angle between the first and second survey lines. He proceeds thus with a third, and so on until all the necessary ground has been covered. If he returns to his starting point it is called a *closed* traverse if he does not it is called an open traverse.

Closed traverse

Closed traverses are those which terminate at the original station point as shown in **1**. A traverse which begins and ends like an open traverse but does so between two points whose positions are known is also considered to be a closed traverse.

Open traverse

Open traverses are those in which measurements of distance and angle cannot be made between the first and last station points. The basic principle is shown in **2**. It is apparent from **1** and **2** that one of the chief differences between closed and open traverses is that the latter produces a polygonal figure and hence it is possible to check results in a manner impossible with an open traverse.

Two checks, one of which is self-evident, are implicit in a closed traverse: coincidence in plotting when returning to the starting station-point and the sum of the angles. The sum of the angles in a polygon is as follows.

(*a*) Sum total of *exterior* angles = 180 (number of sides) deg + 360deg.

(*b*) Sum total of interior angles = 180 (number of sides) deg −360deg.

In the case of open traverses it is sometimes pos-

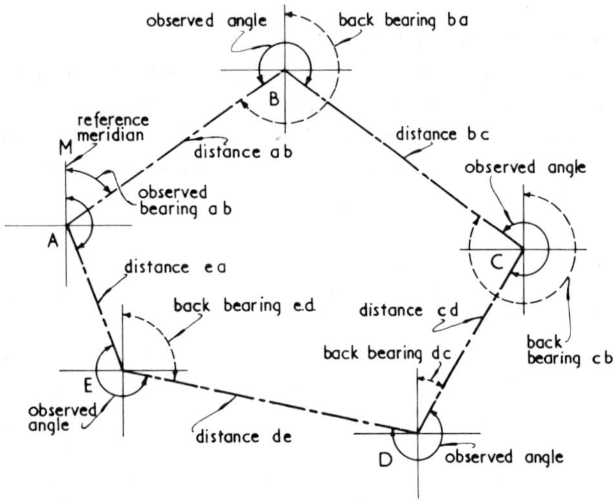

1 *Example showing main lines of a closed traverse*

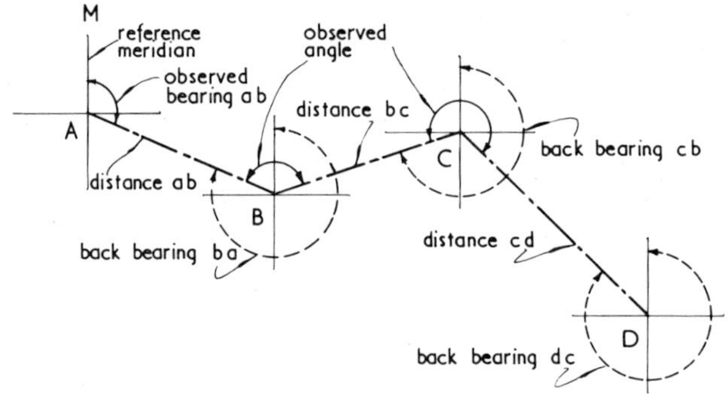

2 *Example showing main lines of an open traverse*

sible to sight back from intermediate or to intermediate station points during the traverse. This should be done where possible as it does provide an angular check. The other check, which should also be carried out is to observe magnetic north at each station point. This provides a check against gross angular errors although it is inadequate for checking small ones. (See Chapters 2.4 and 3.)
Corrections for error are described in Chapter 13. It should be noted, however, that while a level fitted with a horizontal circle or comparable instruments may serve for a closed traverse, a theodolite or its equivalent should be employed for open traverses with the more elaborate of the observational techniques described in Chapter 6.

Linear measurement
Traverse procedure for the measurement of distance is similar to that previously described for chaining and booking. The chief difference is that unlike triangulation it is unnecessary, when traversing, to measure chain lines merely as check lines. All the lines in a traverse can serve a useful purpose for offset measurement.

Measurement of angles

Traverse procedure for the measurement and recording of angles in the field is designed to simplify drawing office work of calculation and plotting. It cannot be too strongly recommended that this should be standardised for any one architect's or surveyor's office and carried out methodically.

The practices recommended in this chapter are not the only ones but have been selected in the belief that they will produce the best all-round results in a wide diversity of offices and circumstances. Standard works of reference should be consulted for further information.

Reference to Chapter 10 and **1** and **2** will show that the angles required for plotting are the bearings relative to the reference meridian. The angles that are observed in the field are, however, those between previous and forward station points (apart from any others that may be required). Bearings must, in consequence, be deduced.

Angles are observed in a clockwise direction

At station points such as B, C and so on in **1** and **2** the principal angle observed with whatever instrument is used is that between a backsight to the previous station and a foresight to the next one. Thus at B; A would be sighted first then C, giving the angle ABC on the instrument.

Angular observations as described above would give *exterior* angles of a polygon if the circuit A, B, C, D and so on were traversed in a clockwise direction as indicated by the arrows or the *included* angles if the traverse were done in the reverse direction (still observing angles clockwise).

Whichever direction is followed it is necessary to translate the observed angles to bearings. The way in which this is done is illustrated in **3**.

The rule is: to find the bearing to a forward station add the observed angle (backsight–foresight) to the backbearing and subtract 360deg if the sum is greater than that.

On smaller surveys some surveyors avoid excessive booking by observing bearings directly. This can be done in two ways. (*a*) By setting the back-bearing for the first (backsight) observation on the horizontal circle of the instrument before observing at each station or (*b*) by transferring the instrument from one station to the next with the foresight reading still on the instrument (it is most important in this case to see that the upper plate clamp remains clamped and that the setting is not disturbed at all) and on arrival transitting (reversing face) the telescope before sighting for the backsight using the lower plate clamp and slow-

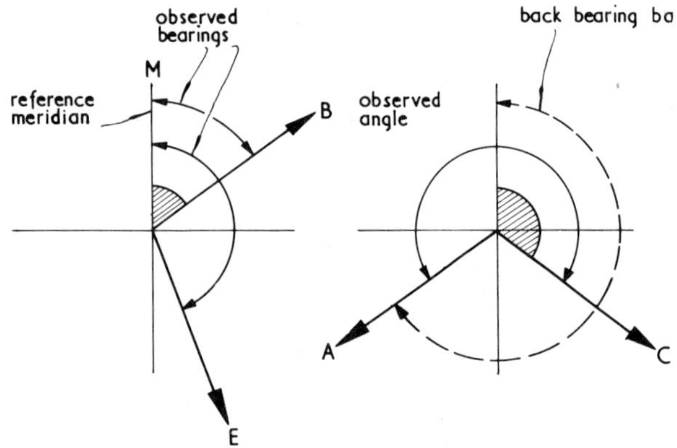

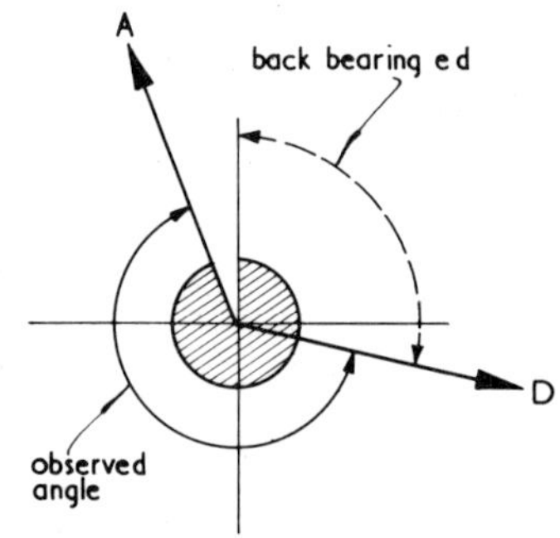

3 Five examples from the closed traverse, to show how whole circle bearings from a reference meridian are obtained from observed whole circle angles by adding the relevant back-bearing and whole circle angle and subtracting 360deg. when the total exceeds that amount. The whole circle bearings are shown shaded

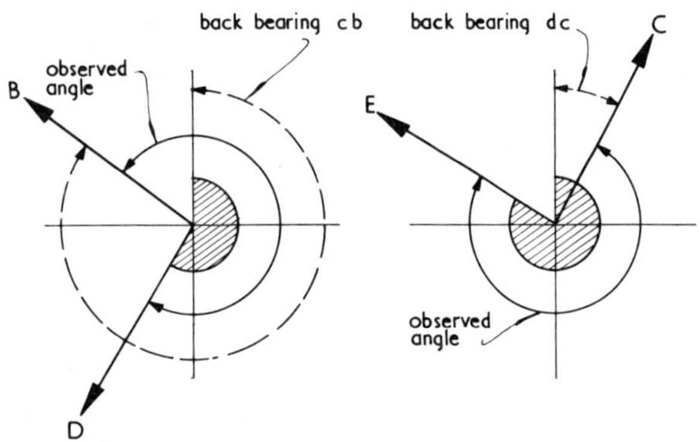

motion screw. This has the same effect as setting the backsight on the instrument but is more fallible.

In both cases (*a*) and (*b*) the result is that the new bearing is observed directly.

Fieldwork

The method of setting out a traverse is the same as that for chaining but it is generally best to make angular observations before any chaining is done.

Chain lines should be decided upon so that they conform as closely as possible to the boundaries of the area to be surveyed, bearing in mind that they should be as long as possible while still permitting reasonable offset measurements.

Wherever possible intermediate angles to stations (other than backsight and foresight angles) visible from each station point visited should be taken so as to provide an extra check. This is especially important when short survey lines cannot be avoided.

Careful centering of the instrument over the station marker is important and station points should be chosen with this in mind.

The theodolite or other instrument used is carried to each station point in turn, set up in the manner described in Chapter 6, adjusted and angles observed and booked.

Angles can be booked in special fieldbooks designed for this purpose, but on relatively small surveys consistent with architectural practice it is better to use the right hand side of an ordinary chaining field-book. Some surveyors prefer this method even for large surveys.

An example of an area surveyed by traversing is illustrated in Chapter 9, **15**.

An example of booking angles etc for a portion of the above example is shown in Chapter 9, **19**.

Reduced bearings

When co-ordinate calculations are being worked out from bearing and distance data it is necessary to find the trigonometrical functions of sine, cosine, and so on of the angles. The majority of books of mathematical tables do not list the value of these functions for angles greater than 90deg. Thus, when bearings exceed 90deg it becomes necessary to *reduce* them to the angular values which give the equivalent trigonometrical functions. The rule is:

If the bearing is between 0deg and 90deg; no reduction is needed.

Ditto 90deg and 180deg; subtract it from 180deg.

Ditto 180deg and 270deg; subtract 180deg from it.

Ditto 270deg and 360deg; subtract it from 360deg.

Study of **3** will show why this works, relative to the co-ordinates of c, for example, which could be calculated in two ways, but for which a reduced bearing has been used.

Chapter 12
Surveying techniques: Levelling

This chapter describes surveying techniques commonly used for determining levels or for establishing contour lines on a site

General

There is no adequate substitute for instruments when it is important to know whether ground is level or to what extent it varies in level. The naked eye is very easily deceived and experienced surveyors know better than to rely on purely visual judgements.

The determination of differences in height is usually termed levelling and contouring. This chapter is concerned only with levelling with the use of instruments, such as levels, telescopic alidades or theodolites used as levels (ie without employing vertical angles).

Datum points

The first step in any levelling operation is to establish a datum point. This is a point to which all other levels must be referred. It may be either a suitable point chosen at the time of the survey or an ordnance survey bench mark.

A local datum point should be an easily identified and permanent spot, preferably at or near ground level. Stone steps, kerbs, manhole covers and the like are usually suitable. Road surfaces are only suitable for temporary reference.

Ordnance survey bench marks

Ordnance survey bench marks are permanent marks cut in walls, on monuments or other locations where a high degree of permanency is to be expected. Ordnance survey bench marks are shown in **1**, **2** and **3**.

Ordnance survey bench marks all over Britain relate to ordnance datum, which is the mean sea level at Newlyn in Cornwall. This datum was adopted in 1921 in preference to mean sea level at Liverpool originally adopted in 1844.

The height of ordnance survey bench marks above ordnance datum are given, together with their approximate location, on ordnance maps. The six-inch and twenty-five inch to the mile maps are being revised in accordance with the Newlyn datum. If the Newlyn datum is applicable to a particular map it will be found to be so noted in the margin.

It is important, especially when more than one map

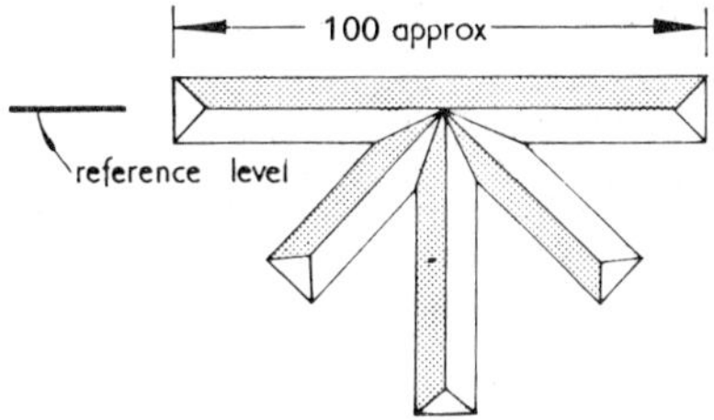

1 *The commonest form of Ordnance bench mark which is usually to be found cut in walls and so on*

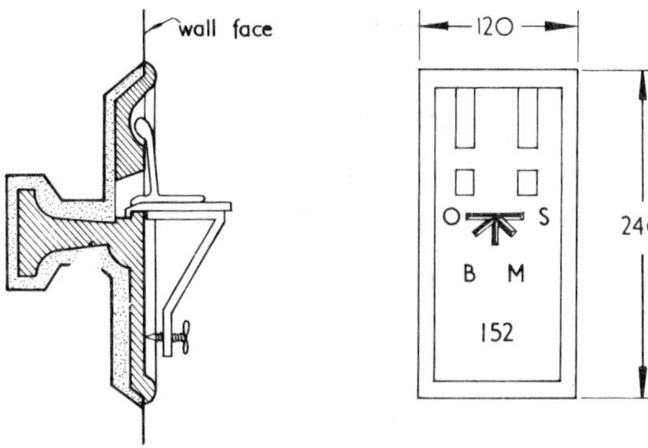

2 *Special form of bench mark less commonly encountered. Section (left) showing special bracket on which staff can be rested; elevation showing slots for bracket*

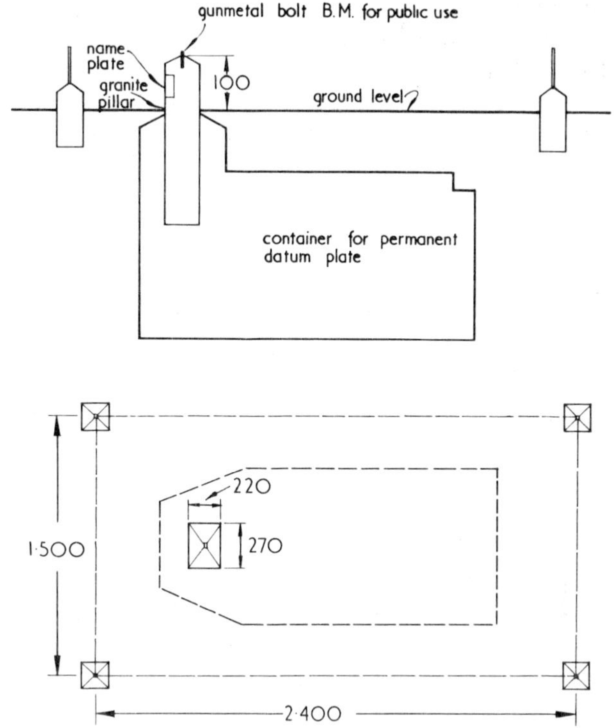

3 *Special permanent key datum point. The permanent reference is kept in an underground chamber but a reference point on a pillar is incorporated for everyday use. Access to the chamber is through a manhole (not shown on drawing) but is not available to the public*

is being used, to see that there is no mix-up between the Newlyn and Liverpool datum values. It is, however, of little practical importance which one is used provided that drawings or records are suitably noted.

Levelling may be a mere comparison of height difference between two points **4**. In practice however it is usually necessary to find differences in level for a number of points.

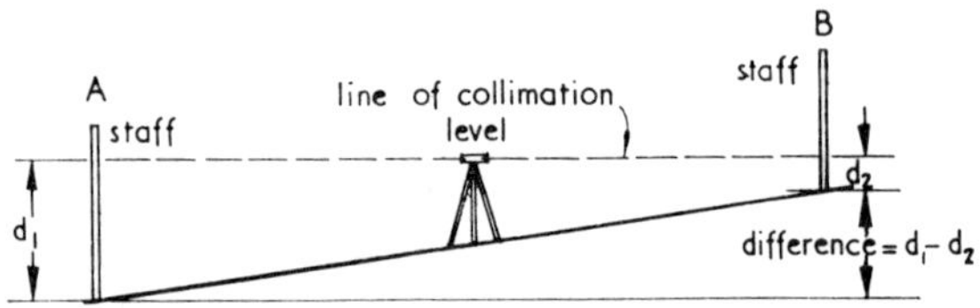

4 *Example of simple levelling to obtain height difference between two points A and B. Difference in level is $d_1 - d_2$*

The first observation in taking levels is to a datum. This is termed taking a *backsight*. The last observation is to a forward point, termed a *foresight*. Observations between these sights are called *intermediate sights*.

Section levelling

Section levelling is the term used to describe the taking of levels along lines such as chain lines so that a section of the land can be drawn.

Section levels comprise backsights, intermediate sights and foresights. The instrument is set up so that as many points as possible can be observed from one station point **5**. A closed circuit is advisable but intermediate sights are taken between each backsight and foresight.

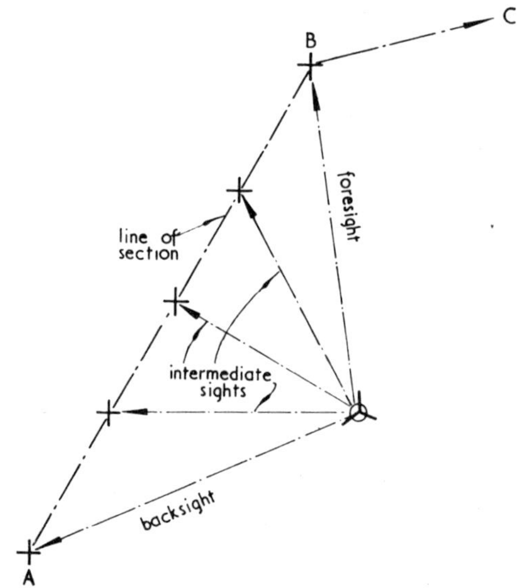

5 *Example of section levelling*

Contouring

Plotting contours (contouring) means drawing lines connecting levels of the same height above datum (as seen on plan). In order to do this it is necessary to take a comparatively large number of levels covering the whole area of the site. Taking levels for contouring follows the same procedure as that for section levelling but with the difference that a pattern of points is observed. The usual patterns employed are rectilinear grids or a series of parallel lines with dimensions chosen for convenience according to the nature of the land to be surveyed, **6**, **7**.

The increased number of intermediate sights characterising contouring increases the probability of uncheckable error but the need for precision is cancelled out to some extent by local irregularities of land surface which can easily produce differences of 150mm or so in a radius of 1m. These are relatively unimportant when discussing the general land form. In the case of hard, made up surfaces the matter is different and at least a proportion of intermediate sights should be made from two instrument station points to points carefully marked on the ground so as to provide a check against gross error.

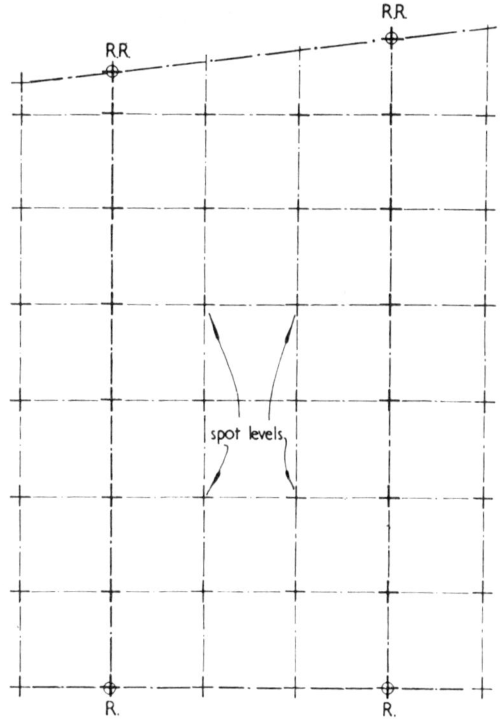

6 *Rectangular grid patterns of levels for establishing contours. Every third line marked out with ranging rods (R to RR) can provide backbones from which subsidiary grid lines can easily be located*

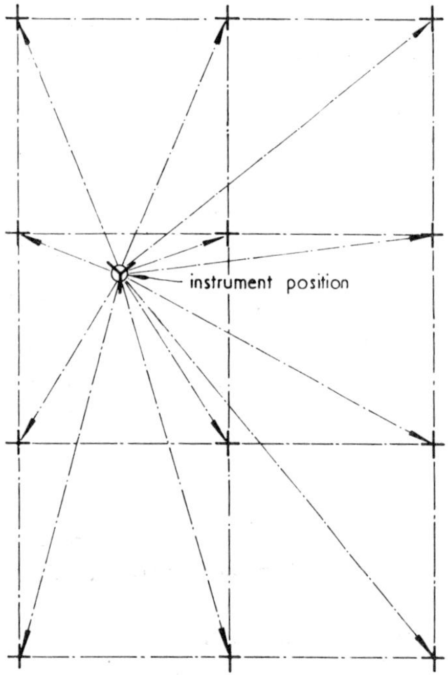

7 Taking levels to establish contours. The instrument should be placed to cover as many points on the contours grid as possible at one time

Flying levels

If the nearest ordnance survey bench mark is too far away to be observed from a site where levelling is to be done and the survey is required to relate to ordnance datum, then *flying levels* must be taken to transfer datum level to a point near the site. This involves taking a series of levels (as few as possible) from the ordnance survey bench mark to the site **8**. Flying levels serve no other purpose.

Obstacles when taking levels

The number of points at which it is possible to read a staff from one instrument station point is limited by a number of factors:

Distance

The quality of the telescope limits the distance at which the staff can be read (See Chapters 6 and 8.)

Slope

Differences in level exceeding 4m (or any other maximum staff length) below the line of collimation cannot be observed nor can differences of approximately 1·5m *above* ground level at the observers' feet.

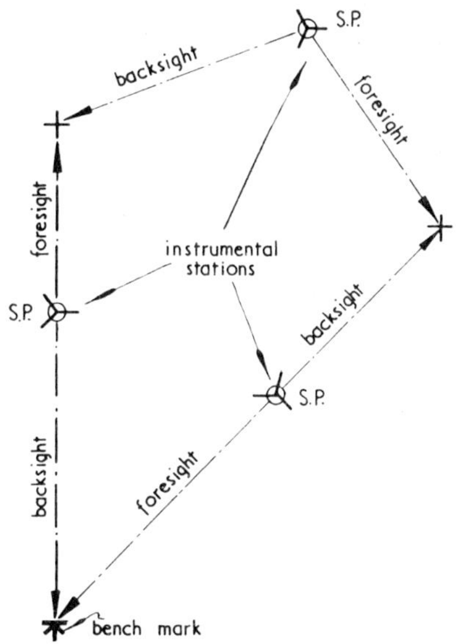

8 Flying levels employ the smallest possible number of observations, preferably in a closed circuit to and from a known datum point. This latter is usually an Ordnance Survey bench mark. The closed circuit acts as a check on accuracy and permits arithmetical correction of minor errors. Alternatively flying levels may be run from one Ordnance Survey bench mark to another with the same object in mind

Obstruction

It is often surprisingly difficult to find an instrument station point, especially in urban areas, from which a convenient backsight and foresight can be made. Trees, buildings, fence posts and similar items also make observation difficult at times.

Levelling procedure

Procedure follows three phases: 1 reconnaissance, 2 establishing station points and 3 making observations. Where larger areas are concerned it may be necessary to follow this procedure for a series of smaller areas. In that case a general reconnaissance is a necessary preliminary and it will probably be advisable to set up change stations from which work on one section can be carried over into the next.

1 Reconnaissance

It is always advisable to walk over ground to be surveyed before beginning operations. As a general rule this will be done as part of the preliminary work for choosing chain lines etc. However, once the terminal points of lines of sections have been decided instrument stations must also be chosen and this usually merits a second reconnaissance with levelling in mind.

2 Establishing station points

All the positions to which backsights and foresights are to be made should be marked with ranging rods (for the day's stint). Instrument station points should then be chosen and (preferably) marked so that there is no doubt about distance, slope and obstruction. Distance will depend upon the magnification of the telescope and weather conditions. A pocket reflecting level is handy for checking for slope. Avoiding obstructions is a matter of common sense.

Positions chosen for instrument station points should be as nearly equidistant as possible from the points to which backsights and foresights are to be made. This is important because it tends to cancel out small instrumental errors (which are unavoidable), and the combined effects of refraction through the atmosphere and the earth's curvature.

3 Making observations

Section levels

The instrument (level, plane table, tacheometer or theodolite) is set up as previously described. When it is ready the staff-man holds the levelling staff up at the datum point. The first backsight is then observed and booked and checked by a second glance through the telescope. If stadia lines are fitted to the instrument it is good practice to record the distance as well because it may act as a guide for the magnitude of the closing errors. Intermediate sights are then taken and, when complete, the staff-man is finally signalled to go to the point to which foresights are to be taken and observations are made and booked as before. Intermediate readings are often booked to the nearest 10mm since accuracy at these points is not usually vital. Where the ground slopes transversely to the section line it is advisable to take *spot levels* either side of it **9**. This is best done an agreed distance (usually paces) at right angles to the section line from the intermediate points.

When an instrument has a horizontal circle, or when a telescope alidade is being used, spot levels are much easier since distance (by stadia intercept) and horizontal angle from the instrument station point can be read off by the observer. It is, of course, necessary to record the position of the level and this can be done by the same means relative to the backsight and foresight points for each line of sections, **10**.

Contouring

The procedure commonly followed is to lay out a series of parallel section lines over the area to be

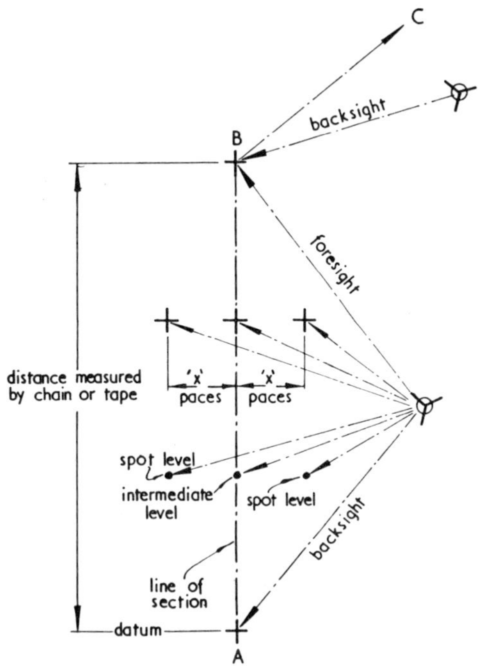

9 *Section levels on land with a transverse slope may require spot levels to be observed on either side of the main chain line (line of section)*

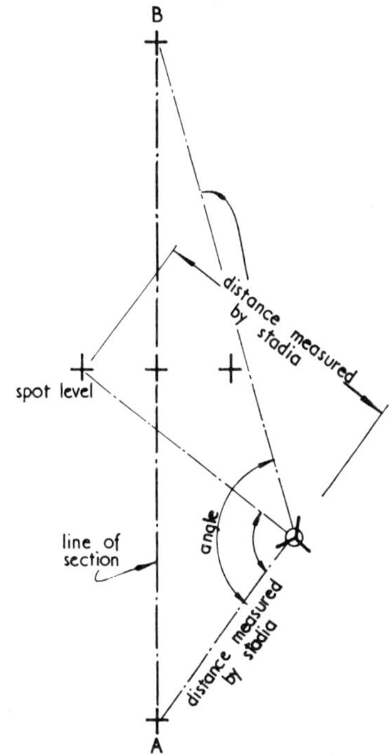

10 *The position of the level or location of spot level points may be determined by stadia readings*

surveyed. The corners of the survey area are marked with ranging rods. Pegs are then driven in at regular intervals along the lines between ranging rods on two opposed boundaries, **6**. The instrument should be set up to cover as much of the area as possible.

The whole procedure is much simplified if stadia observations are possible and a plane table with telescopic alidade is probably the ideal instrument to use (in fine weather). It is usually possible to observe several section lines from one position of the instrument. If stadia observation is not possible, or sometimes from preference, a chain or tape can be used between ranging rods with the staff man instructed to set up the staff at suitable intervals.

Flying levels

Procedure for taking flying levels is the same as for section levelling except that intermediate sights are not required.

Booking levels

There are several methods of booking readings during levelling. Two of the most usual are called the *collimation method* and the *rise and fall method*. The principles upon which each method is based are illustrated in **13** and **14**.

Level books may be purchased with rulings suitable for recording by the collimation method and by the rise and fall method, **15** and **16**.

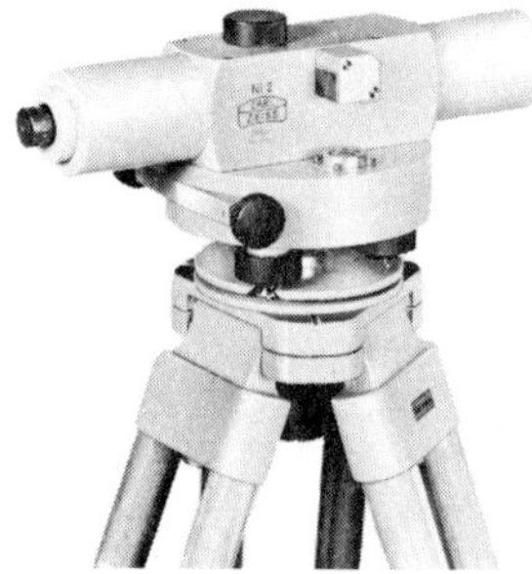

11 *The Carl Zeiss Ni 2 is an example of one of the first of the new kinds of level that have been developed by leading manufacturers; after an easy initial adjustment with a circular bubble the level sets itself so that it can be used with a high degree of accuracy. Speed of operation is almost double that of conventional levels*

12 *The Carl Zeiss Ni 4 illustrates a trend similar to the Ni 2 but aimed at the construction industry. These instruments are sturdy, accurate and admirably suited for work on building sites. This particular instrument is unique in that a circular bubble inside it is used as a lens in the automatic self-levelling optical system*

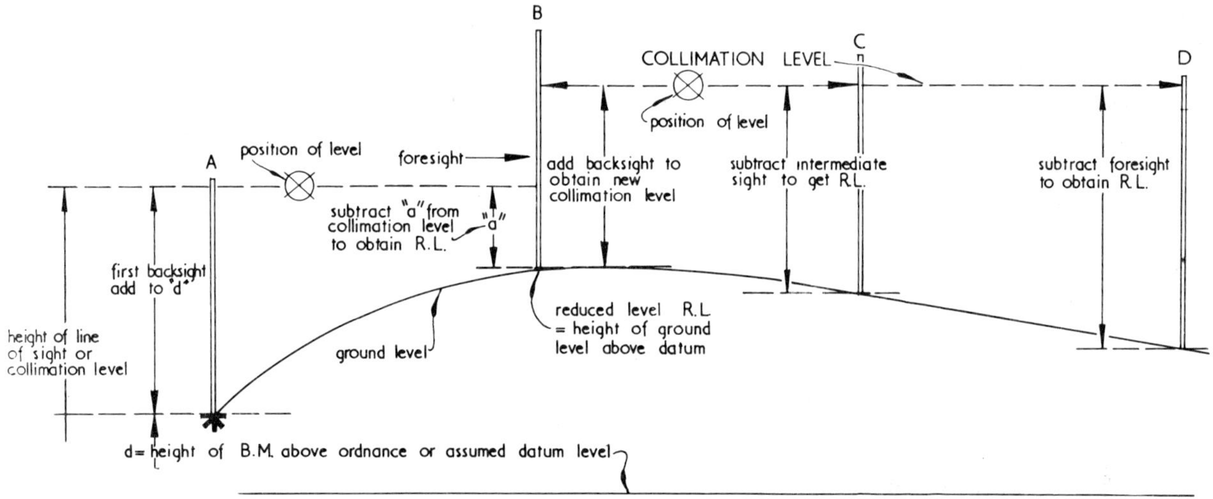

13 *The collimation or height of instrument method of recording levels*

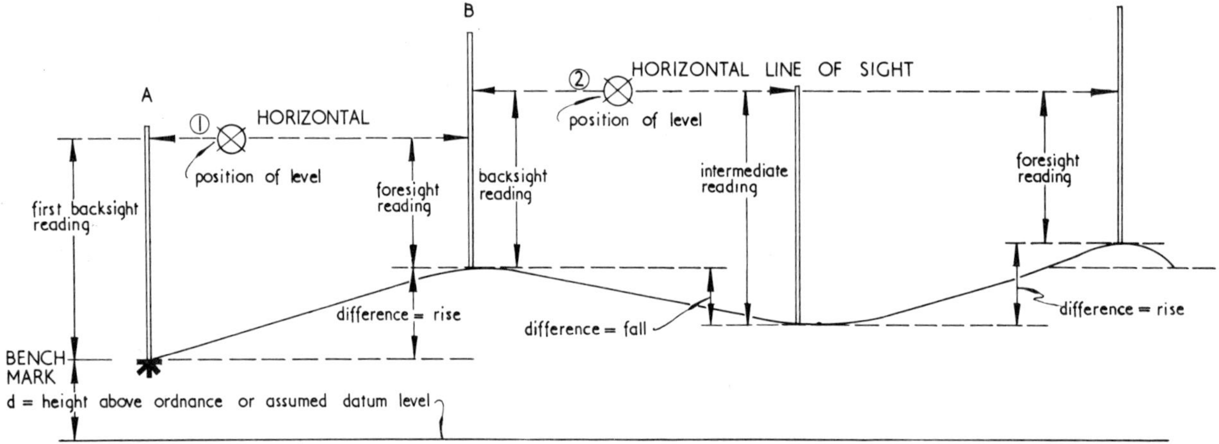

14 *The 'rise and fall' method of recording levels*

Descriptive title of work recorded.

Backsight	Intermediate	Foresight	Instrument height	Reduced levels	Distance	Remarks

Enter:– Backsight, then Intermediates then the foresight. The next backsight is put on the same line as the instrument height for the same position; at change points the sequence is then repeated.

Enter: a brief description of each position of the staff

Enter: distance of staff from level by stadia reading, or for section levels, distance along chain line

Enter: height of instrument minus foresight, intermediate or backsight as relevant.

Enter: first backsight added to datum level to obtain instrument height and subsequently backsights to corresponding reduced levels to obtain instrument height i.e: height of line of collimation

EXAMPLE: Collimation Method.

Backsight	Intermediate	Foresight	Instrument height	Reduced levels	Distance	Remarks
3.872			36.018	32.146	5	O.D. on church porch. = 32.146
2.402		0.733	37.687	35.285	29	Station A: road mark.
	1.145			36.542	44	Step of No3, High St.
0.326		1.782	36.231	35.905	48	Station B M/H cover.
0.765		2.128	34.868	34.103	55	Station C. road mark.
2.003		0.658	36.213	34.210	60.5	Station D. road mark.
		0.962		35.251	72	OD. on P.O. wall = 35.251
4.368		6.263				

Method for checking: Sum of foresights = 6.263
Sum of backsights = 9.368
Difference = 3.105

The two differences should be the same.
First reduced level = 32.146
Last reduced level = 35.251
Difference = 3.105

The collimation method is best understood by doing it. The following figures work out correctly.

1.672						O.D = 24.357
		3.861				
0.344						
	2.206	2.748				
1.202						
	0.989	3.445				O.D = 17.521

15 *Booking levels for the collimation or height of instrument*

Descriptive title of work recorded

Backsight	Intermediate	Foresight	Rise	Fall	Reduced levels	Distance	Remarks

Enter:- Backsight, then Intermediates then the foresight. The next backsight is put on the same line as the foresight to the same position; at change points and the sequence is then repeated.

Enter: a brief description of each position of the staff

Enter: distance of staff from level by stadia reading or, for section levels, distance along chain line.

Enter: datum level and subsequent additions of Rise, or subtractions of fall from the preceding figure.

Enter: rise or fall as the case may be; obtained by finding the differences between backsight and relevant level staff readings. The sequence begins anew with each backsight

EXAMPLE: Rise and Fall Method

Backsight	Intermediate	Foresight	Rise	Fall	Reduced levels	Distance	Remarks
3.872					34.146	5	O.D on church porch = 32.146
2.402		0.733	3.139		35.285	29	Station A. road mark
	1.145		1.257		36.542	44	step of No 3 High St.
0.326		1.782		0.637	35.905	48	Station B. M/H cover
0.765		2.128		1.802	34.103	55	Station C. road mark.
2.003		0.658	0.107		34.210	60.5	Station D. road mark
		0.962	1.041		35.251	72	O.D. on P.O. wall
9.368		6.263	5.544	2.439			R/F difference =

Method for checking: Sum of foresights = 6.263
Sum of backsights = 9.368
difference = 3.105

The three differences should be the same.
Difference between sums of rise and fall: 3.105
Difference between first and last reduced levels: 3.105

The rise and fall method is best understood by doing it: The following figures work out correctly:

1.672							O.D. 24.357
0.344		3.861					
	2.206						
1.202		2.748					
	0.989						
		3.445					O.D. 17.521

16 *Booking levels for 'rise and fall' method*

Chapter 13

Surveying techniques: Drawing office work

Since drawing office practices and standards for surveying are different from those required for architectural work, this chapter recommends methods appropriate for survey work with some notes on the treatment of unavoidable errors. It does not include data for trigonometry, mensuration, geometry and basic drawing techniques for which appropriate text-books should be consulted

General

Standards of draughtsmanship in architects' offices are normally inappropriate for survey drawing, for which a high standard of accuracy is needed. Thus when plotting surveys, especially when wholly graphical methods are used, most architects and draughtsmen accustomed to architectural work need to re-think their technique for the occasion.

The chief reason for a lack of precision in architectural drawing are as follows:

1 drawing instruments are usually slightly inaccurate;

2 drawing paper, linen and tracing papers in normal use are dimensionally unstable;

3 considerable reliance is placed upon written or calculated dimensions;

4 tolerances normal in building work make very precise draughtsmanship unnecessary;

5 excessive precision is usually uneconomic.

When preparing survey drawings the points given above need consideration although the importance attached to each will vary with circumstance.

Checking drawing instruments

Tee-squares Draw a line right across the paper on a drawing board with the tee-square. Turn the tee-square end to end and lay it along the line previously drawn. The checkline drawn with the tee-square will probably show a slight bowing. This may be because the tee-square is not straight but is more likely to be because the angle at which the pencil was held was not constant.

Set squares To check a set-square (if the tee-square is all right) draw a vertical with it, reverse the set-square and see if it corresponds exactly with the line previously drawn. Set-squares are seldom truly right-angular and any important right angles, such as reference meridian lines should be set up geometrically with a beam-compass. It is also important to realise that drawing boards are almost never truly square and the practice of using an edge other than the ebony edge normally leads to inaccuracy.

Preparing drawing materials

Dimensional instability of drawing paper or linen matters if any reliance is going to be placed upon scaled dimensions or graphical methods of plotting. Stretched or mounted good quality drawing paper (eg Whatmans) should be cut off the board and kept flat in a drawer in the office in which it is to be used for about a week before use. Drawing on a mounted sheet is satisfactory provided that work is complete; all dimensions are recorded numerically; and no reliance is placed upon unchecked important scaled dimensions after it has been cut off. This is because it will have contracted. Sheets will still expand or contract, but fairly evenly, once they have been cut off—hence the practice of drawing a scale of some length. An alternative to good quality drawing paper is mill-board, which should be clipped but not gummed or glued down. Mill-board, preferably grey to reduce glare, is especially appropriate for plane-table work. Linen is probably the best material for tracing survey drawings although some of the new transparent plastics are reasonably satisfactory. Both kinds of material should be cut into sheets of appropriate size and hung from clips for a week or more before use. Prints of survey drawings should be 'true-to-scale' since ferro-gallic and dye-line prints distort to some extent, usually unevenly as regards length compared with width.

Survey drawings

The mental attitudes of surveying and architectural draughtsmen are fundamentally different since the former uses drawing to discover (unmeasured) dimensions and delineate what exists while the latter uses drawing to describe dimensions that arise from design decisions. The co-ordinate system of plotting (see Chapter 10) is especially appropriate for architectural survey drawings because co-ordinate-grids provide a dimensional framework within which small errors due to draughting are localised and relatively unimportant.

Draughtsmanship

Precise draughtsmanship demands refinements such as those practised by some surveyors who, when laying out the main lines of a survey, use a steel straight-edge in preference to a tee-square and draw lines with the eye-end of a needle; which indents a fine line without scratching the paper.

Fine needle-points are used to prick-off station points, and also convenient chainage lengths (eg 10m, 20m) along chain lines. Beam compasses are used for intersecting arcs. Detail other than the main lines of a survey is put in with a hard fine-pointed pencil. When a survey has been drawn—at least as regards its main parts—and 'proved' it can be lined-in with thin ink lines or a medium pencil sharpened to a chisel point.

Drawing costs

Architectural draughtsmanship can become unnecessarily expensive if it is over-meticulous. Good survey draughtsmanship however, with its accurate if time-consuming carefulness, at least obviates the kind of very expensive mistakes that can arise when a site turns out to be even a little different from the drawing of it used to prepare working drawings.

Plotting surveys

Plotting co-ordinates

To prepare a sheet for survey plotting by rectangular co-ordinates first draw a square grid. The dimensions of the grid will vary according to the scale to be used but it is best if the squares have sides of between 50mm and 150mm. This localises the effects of expansion and contraction satisfactorily. Great care should be taken to ensure accuracy at this stage, using a needle point to prick dimensions and very fine lines. (See Chapter 10.)

Plotting chain lines

When plotting a survey the longest line should be drawn first. Procedure varies slightly as between co-ordinate plotting and graphical plotting.

Setting out from co-ordinates, for which methods of calculation are given in Chapter 10, diagram **3** is done by measuring northings up from the nearest latitude grid line and to the right from the nearest meridian grid line and then drawing fine lines parallel to each. In both cases it will of course be necessary to deduct the value of the grid line from the co-ordinate numbers. Thus if the co-ordinates were 516.4E 262.3N and the nearest grid lines represented 500 East and 250 North; the measured distances would be 16·4 perpendicularly to the right and 12·3 perpendicularly up from the 500E and 250N grid lines respectively. The intersection of the two fine parallel lines would then locate the point required. When the first survey line has been drawn, points of intersection with other lines are then plotted along it—either from calculated co-ordinates or by scaling as more appropriate. All the main lines of the survey should be completed in this way before plotting detail.

Graphical plotting

In the case of graphical plotting the longest line should be drawn and its length carefully marked off with a scale; the station-points being pricked in and ringed. Triangles extending from the base line are next constructed with the lengths of sides as radii, using a beam compass and a horn centre under the compass point. (See Fraser Reekie, R. *Draughtsmanship*, Edward Arnold & Co). When the main triangles have been checked by means of the check lines (Chapter 9) and the main survey framework completed smaller triangles may then be drawn in.

Plotting a traverse survey

When traverses are plotted graphically accuracy in measurement has to be extended to accuracy in setting out angles. The first line is drawn as previously described and then the necessary angle, made with it by the second line, must be set out. This can be done by one of two methods:

1 *By protractor* This is the quickest method and satisfactory for most purposes provided that a suitable protractor is used. Ordinary protractors of about 150mm diameter only permit plotting to within about 10 or 15 minutes which is inadequate if a good instrument has been used for the survey. Elaborate protractors of 200mm to 300mm diameter and fitted with verniers and a pricking needle can be used, but are often difficult to centre accurately. Protractors of 300mm to 450mm diameter, with graduations printed on cardboard, are inexpensive and probably the best suited for protracting angles.

2 *By the tangent method* This is explained in **1**. With care the tangent method produces more accurate results than the use of a protractor. The lines forming station points and tangent points should be as long as possible. The tangent method is based on the fact that the tangent of an angle equals the ratio opposite side of triangle to adjacent side in a right angled triangle.

Detail work

Plotting detail is almost wholly a matter of drawing from offsets recorded in the field-book. It can be done with an adjustable set square set to give perpendiculars for relevant chain lines and an ordinary scale, or by means of a special offset scale. When an adjustable set square is used each chain line should be marked off with the offset points.

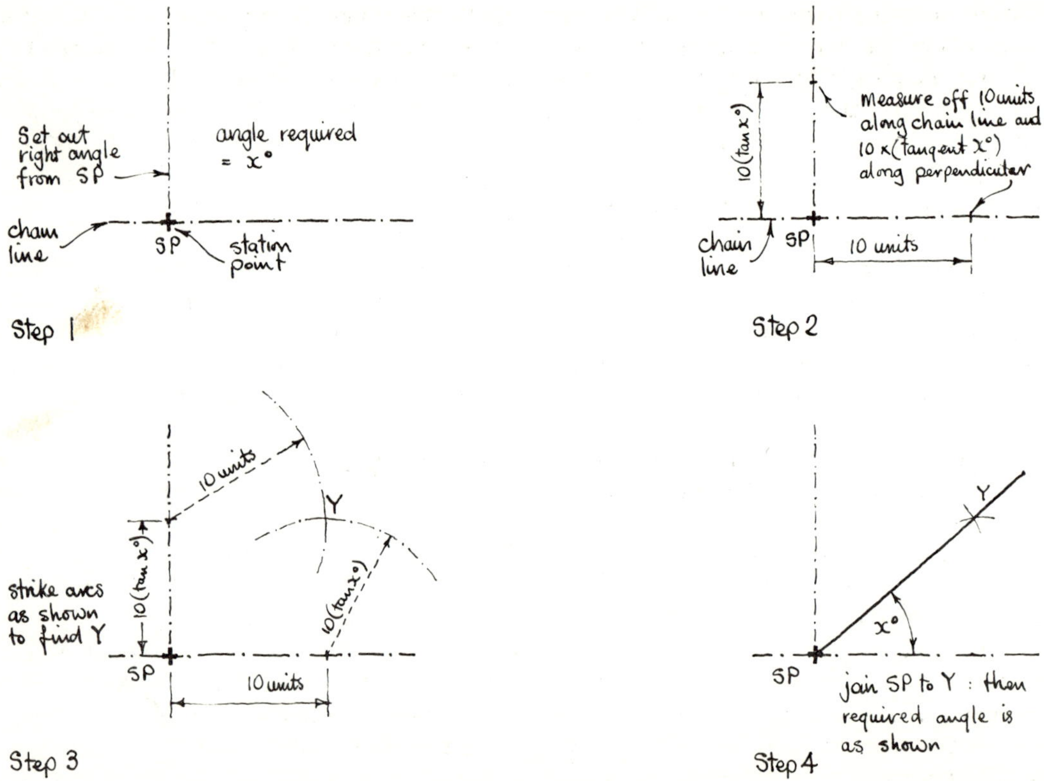

 1 *Plotting angles on survey drawings by the tangent method*

Then faint lines perpendicular to the chain line should be drawn through each extending, to scale, about 6 to 8m either side. Starting at the bottom of the chain line, as recorded in the field book, each offset should then be scaled off and marked with a needle prick or small pencil dot. The dots marking continuous lines (eg irregular boundaries) should be connected with their pencil lines as the work proceeds. This is much less troublesome than attempting to join up all the dots later on.

Offset scales are short broad scales graduated in both directions outward from zero at the centre. They are used by sliding them like a set square against another, long, scale which is placed parallel to the survey line on which offsets are being plotted and so that the zero mark of the offset scale can move along the survey line. Thus lengths along the survey line can be read from the long scale and offset distances, either side of the survey line, from the offset scale.

Some of the mistakes most commonly made in plotting offsets are:

plotting on the wrong side of the survey line;

joining up the wrong points;

scaling offset positions from the wrong end of the chain line.

Conventional symbols greatly simplify description of features commonly occurring. A knowledge of them also makes it much easier to read Ordnance Surveys plans and maps. They are given in the margin of one-inch Ordnance Survey maps and in most text-books from which a selection is given in **2**.

Plotting levels

Generally the initial work of computation of levels will have been done in the field book. In some cases there is little more to do; in others further work will have to be done to produce the necessary information. Three main categories of information may be involved:

levels at specific points, eg invert levels;

the level configuration along strips of land, eg roads or boundaries;

the generaly level configuration or contours of an area of land, eg a field.

Spot levels

Levels will normally have been taken for well defined positions and little remains but to annotate drawings accordingly.

Section levels

Level configuration along strips of land will norm-

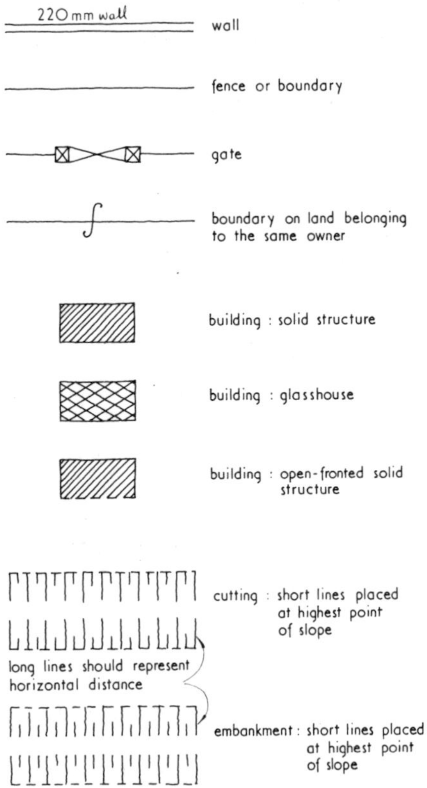

 2 *Conventional symbols for survey drawings*

ally require that levels at points other than those at which the level staff was placed must be found. This is done by taking levels along a survey (chain) line and then plotting a section. To plot the section a straight line is drawn to represent a convenient height above (or below) datum level eg 5m. Perpendiculars are then erected along this line at intervals which correspond, to scale, with the chainage (linear measurement) figures for staff positions as recorded in the field book. The levels from the field book, minus the assumed height of the reference datum line, are then scaled up the relevant perpendiculars to give a series of points which, when joined, represent the levels section. It is common and useful practice to plot the verticals at three or four times the scale of the linear distances. This makes interpolation easier and more accurate.

In some instances the land may slope transversely to the survey line as well as along it. When this is important spot levels (Chapter 12) should have been taken as offsets along the chain line at right angles to the various staff positions on it. If so it makes for clarity to extend the perpendiculars on the relative section line and draw cross-sections on them.

Contouring levels will have been taken either on a 'grid' pattern (Chapter 12) or more or less at random.

In either case a series of spot levels will have been found whose position and value can be plotted in plan. The basic problem in either case will be one of interpolation to find the contours. Contours are lines, joining points of equal level on plan and drawn to represent vertical intervals which are chosen according to the nature of the land: relatively flat land may require a 100mm contour interval but a 200mm interval is usually enough in architectural work. The method of interpolation which is probably the most suitable, whether 'grid pattern' or spot observations have been used, is estimation by eye along lines joining the points of known height. There are other methods but it is doubtful if any resultant increase in accuracy would be worth the disproportionate increase in labour involved for work of a kind encountered in ordinary architectural practice. Examples of contouring with a grid layout and spot level survey of the sort associated with tacheometrical methods are given in **3**.

Permissible errors

The theory of errors is too complex for adequate treatment in an introductory book. At the same time it is not possible to generalise usefully about the application of theory. The notes that follow must, in consequence, be regarded as comments for guidance on small scale surveys rather than authoritative statements. For the latter Clark, D. *Plane and geodetic surveying*, vols I and II, Constable & Co Ltd should be consulted.

Errors and corrections

Chain surveys There is no way to check the results of a chain survey for error if the same chains, etc have been used throughout and are incorrect, except by reference to accurately located independently known points within the survey. The avoidance of and sources of error from linear measurement have been discussed in Chapters 3 and 4.

Plotting errors may arise from limits set by accuracy of draughtsmanship. A good draughtsman can be expected to work to an accuracy of plotting of ¼mm. Thus this amount represents a limit below which distances cannot be represented and hence the limits within which they should be measured. The effect will vary according to the scale at which a survey is to be plotted. The following table gives an indication of the kind of accuracy which should be attempted.

Gross errors in triangulation may be revealed by check lines (see Chapter 9). In that case the only sensible means of checking is to try to discover by trial and error if a chain length has been omitted

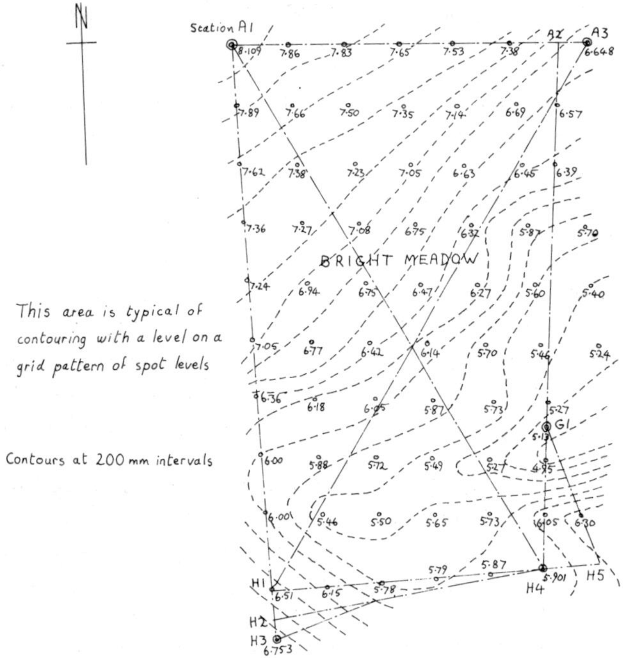

3 *An example of contouring by interpolation from a grid layout and spot levels*

by accident—followed by a quick check in the field by remeasurement of any suspect line thus revealed. Otherwise remeasurement is the only answer, checking against the field-book.

Scale	**Minimum length that can reasonably be represented**
1/10 000	3m
1/5000	1m
1/2500, 1/2000	500mm
1/1250, 1/1000	250mm
1/500	100mm
1/250, 1/200	50mm
1/100	25mm
6in to 1 mile	10ft
1in to 100ft	1ft
1in to 32ft	4in
1in to 20ft	2in
1in to 10ft, 1in to 8ft	1in

Traverses

Traverse errors are a combination of linear measurements and angular errors. The presence of angular error can be detected in a closed traverse by means given in Chapters 6 and 11. The extent of permissible error can be judged by the limits within which it is possible to read the instrument used, eg with an instrument with which it is only possible to read ± 30 seconds a cumulative error proportional to that amount can be expected. The sum of the errors will appear as a 'closing error'. If this does not exceed 1 in 1000 for modestly accurate work, 1 in 2000 for careful work and 1 in 5000 for high grade surveys (excluding first order work such as a geodetic survey) the final error may reasonably be adjusted. One method of adjustment for work with an error of up to 1 in 2000 is that of proportionate allocation.

Proportionate allocation If the closing error in eastings is xm and in northings ym then the total error may be proportioned out for each station point (at which changes of direction occur) according to the length of traverse survey lines. The correction for each station point is obtained as follows:

$$\frac{\text{Length of traverse survey line}}{\text{Total linear distance traversed}}\, x = \begin{array}{l}\text{correction for}\\ \text{eastings}\end{array}$$

$$\frac{\text{Length of traverse survey line}}{\text{Total linear distance traversed}}\, y = \begin{array}{l}\text{correction for}\\ \text{northings}\end{array}$$

Thus, for example, if the closing error were 0·6m E 0·2m N and the total length of traverse 2000m, the correction for a survey line of 500m would be:

$$\frac{500}{2000} \times 0{\cdot}6 = 0{\cdot}15\text{m E} \quad \frac{500}{2000} \times 0{\cdot}2 = 0{\cdot}05\text{m N}$$

Levelling

Checks for error and limits of accuracy are given in Chapters 4 and 6. Some error is to be expected, and provided it does not fall outside acceptable limits can be judiciously distributed between station points. The distribution of error is made proportionately to distance between change points and the level. This is why it is important to measure or record stadia readings for back and foresights.

If the total levelling error is x and the sum of the distances ($l_1 + l_2$ etc) from staff positions to the level is S then the error may be divided out proportionately as follows:

$$\frac{l_1 + l_2}{S}\, x = \text{corrections for station 2.}$$

$$\frac{l_3 + l_4}{S} x = \text{correction for station 3.}$$

and so on.

The distance is given as $l_1 + l_2$ because it represents the distance between staff positions *via the level*: l_1 is the distance between the datum backsight staff position and the level. l_2 is the distance between the level and the foresight staff position.

Chapter 14
Surveying techniques: Setting out buildings

This chapter describes some of the methods commonly used for setting-out buildings. It indicates the kind of information that an architect must supply to the contractor under clause 5 *of the* RIBA Agreement and schedule of conditions of building contract.

Responsibility

Contract agreement The 1963 *Agreement and schedule of conditions of building contract* published by the RIBA contains the following clause relative to setting-out. 'Clause 5. The architect shall determine any levels which may be required for the execution of the works, and shall furnish to the contractor by way of accurately dimensioned drawings such information as shall enable the contractor to set out the works at ground level. Unless the architect shall otherwise instruct, in which case the contract sum shall be adjusted accordingly, the contractor shall be responsible for and shall entirely at his own cost amend any errors arising from his own setting out'.

Site dimensions The first part of clause 5 places the onus for making an accurate site survey, as regards site plan dimensions, upon the architect. The second part limits the contractor's liability to any errors which can be interpreted as or shown to be a failure to follow the architect's instructions for setting-out at or below ground level. The inference is very clear that any errors which can be shown to have arisen from mistakes in the information supplied by the architect will become the architect's liability. It is good practice to obtain the contractor's agreement in writing that adequate setting-out information has been given to him and to agree tolerances.

Levels In clause 5 the phrase 'determine any levels . . .' could be interpreted to mean that the architect must not only fix all the necessary levels of floors, pavings, drains, road surfaces etc, as a matter of design but also take them during operations. Specifications should make it clear that the contractor will be responsible for actually taking levels during building operations. This does not relieve the architect from the responsibility of initially defining those levels by means of ordnance datum or a local datum set up by himself.

Checking setting-out It still remains the architect's responsibility, in his capacity as the building employer's agent, to see that the contractor has not made any errors. It follows that a physical check must be made either by the architect or a clerk of works at various stages throughout the contract. It is especially important to make a stringent check on the main original setting-out. This in turn requires that the architect shall be familiar with setting-out techniques and shall agree levels and setting-out with the contractor at the start of the work.

Tolerances The precision with which points are fixed in plan and section will vary according to the nature of a building and its site. The usual tolerance in dimensions is a maximum error of 5mm in 50m. When extreme accuracy is required tolerances should be decided in advance of contract (probably in consultation with specialists), included in the specification, where relevant, and brought to the contractor's notice.

Equipment

The equipment normally required for setting out should include a steel tape, linen tapes, folding rods, a mechanics level, an optical level, a supply of clean square wooden pegs (about 50 × 50 × 600mm), strong line, a number of rails (about 150mm × 30mm × 1·5 to 2m long), wire nails and boning rods (described later). On complicated or large sites a theodolite should be added to the list, and when tall buildings or deep excavations are concerned an optical plummet will be needed.

Instruments may be provided by the architect or contractor. If they are provided by the contractor the architect should satisfy himself as to their accuracy before they are used for checking.

Preliminary work

Site plan survey The site should be surveyed by whatever method seems appropriate. The survey should have included at least *three*, and preferably more, permanent points—ie points which cannot be moved and which can be precisely identified with certainty. These points, with their reference co-ordinates, should appear on site layout drawings, ground floor plans and any basement plans. Their function is to provide key points for setting-out and an initial check on survey accuracy before setting-out.

Site levels datum At the time of the initial survey levels should have been related either to Ordnance datum (with a clear distinction as to whether the reference was Newlyn or Liverpool) or to an arbitrary and permanent datum on or near the site. Subsequently, if Ordnance datum has been used, a site datum should be set up in the form of a post, concreted in, and located so that it can remain undisturbed throughout building operations. It is remarkable how often an unsatisfactory position is chosen, so the contractor should be consulted about location of stores, site circulation, temporary offices etc, to make sure of freedom from interference or subsidence.

Site level configuration Whether or not the site has been contoured on an original survey it should be relevelled with reference to site datum on a grid of 5m to 10m squares. This should be done by the architect or clerk of works or by the contractor under the architect's supervision. The levels should then be recorded on a site plan **1** and agreed with the contractor. This precludes any arguments as to amounts of cut and fill for which claims might be made.

Survey check The point-to-point distance of survey reference points should be calculated or found from the survey data and measured prior to setting out, using a steel tape. This is to protect the architect from the effects of survey errors or any tampering that might have taken place between the time of the initial survey and measurement of building work.

Baseline Once the original survey reference points have been checked and found correct, or corrected, as the case may be, a baseline must be established. This should be as long as possible within the limits of the site and chosen, preferably, so that it runs parallel to at least one side of the proposed building. The co-ordinates of its extremities should be calculated if the site was plotted on the co-ordinate system. If some other method of plotting was used then the baseline can conveniently be regarded as running parallel with an arbitrary grid meridian and its terminal co-ordinates worked out accordingly. When this has been done the baseline can be set up by very careful measurement with a steel tape from the original survey reference points, or, if this degree of precision is unnecessary, by reference to a convenient boundary. The baseline terminals **2** should be marked by posts driven well into the ground and having wire nails or long panel pins driven in their heads to mark the point of measurement. In some instances setting up a secondary baseline or reference latitude at right angles to the first is sound procedure.

Setting out drawings Drawings should be prepared for setting-out in accordance with clause 5, either by making a separate drawing or by annotating working drawings. The method chosen should accord with the complexity of the work but in all cases should refer back to the baseline either by

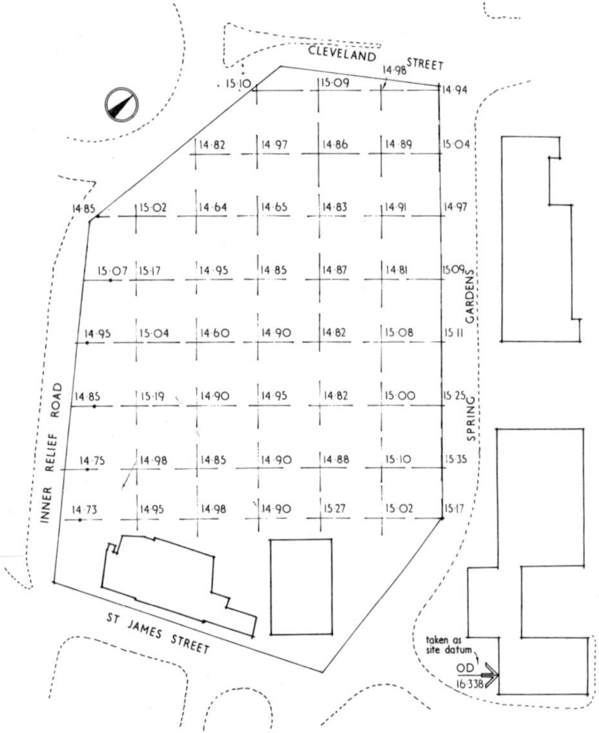

1 *Diagram is a reproduction of site plan levelled by contractor and agreed with architect. Site datum in this instance was a convenient ordnance datum*

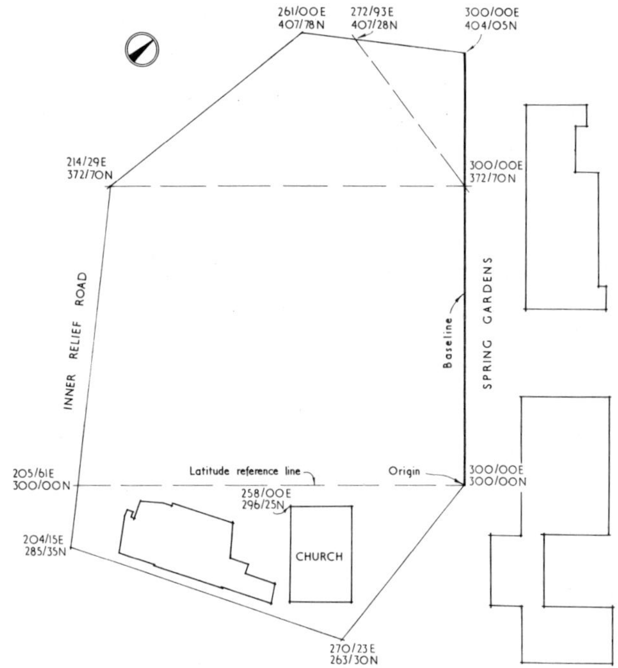

2 *Diagram illustrates same site as 1 with key survey co-ordinates and baseline used for subsequent drawings marked on*

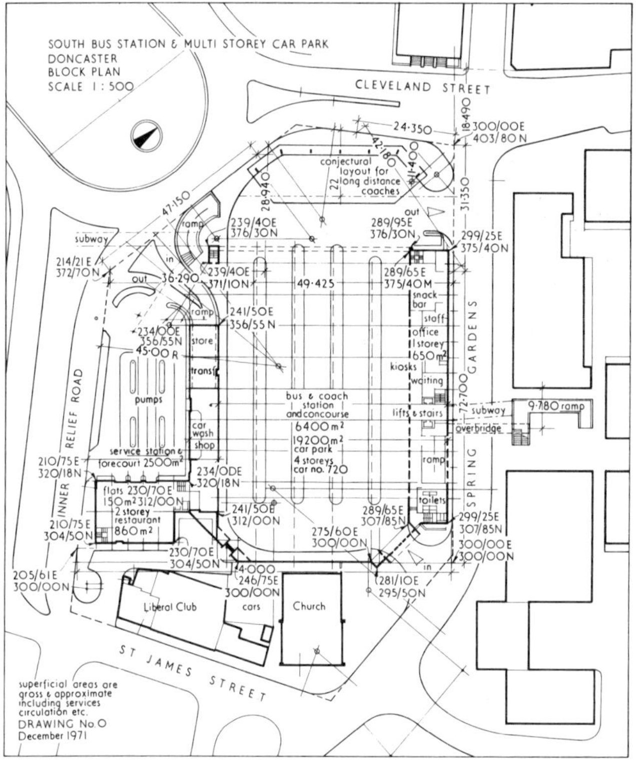

3 *Diagram is a reproduction of site plan from which 2 was prepared. It shows key dimensions, main co-ordinates and other essential information for setting out*

written dimensions **3** and setting-out lines (ie lines along which dimensions are to be measured) or by co-ordinates.

Setting-out co-ordinates The advantages of the co-ordinate system become especially apparent when dealing with large or complex buildings: it is necessary only to annotate key points on the working drawings with their co-ordinates to fix their location and check against gross error by careful scaling. Not only does this enable a great many more key points to be fixed accurately without multiplying confusing lines but it also enables larger scale drawings and details to be annotated similarly, and accurately locates column or stanchion centres, service outlets and similar features. Additionally it greatly simplifies the determination of curves of large radius or irregular form. Examples of setting-out drawings are given in **4** and **5**.

Setting-out

Plan dimensions The first stage in setting-out on site is to drive pegs in at the key points—eg the four corners of a rectangular building—by means of measurement from the baseline and/or reference latitude line. Alternatively when convenient to do so, points can be located by triangulation or co-

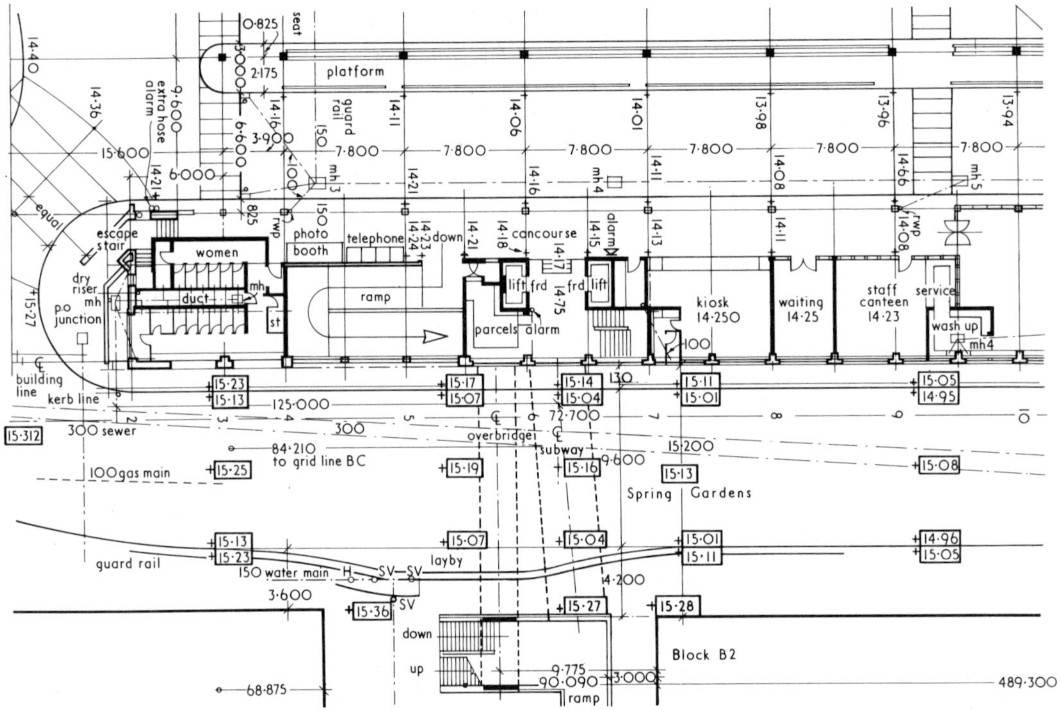

4 *Diagram shows a portion of ground floor plan of working drawings for building shown in 3. It gives levels and dimensions within framework of main setting-out co-ordinates*

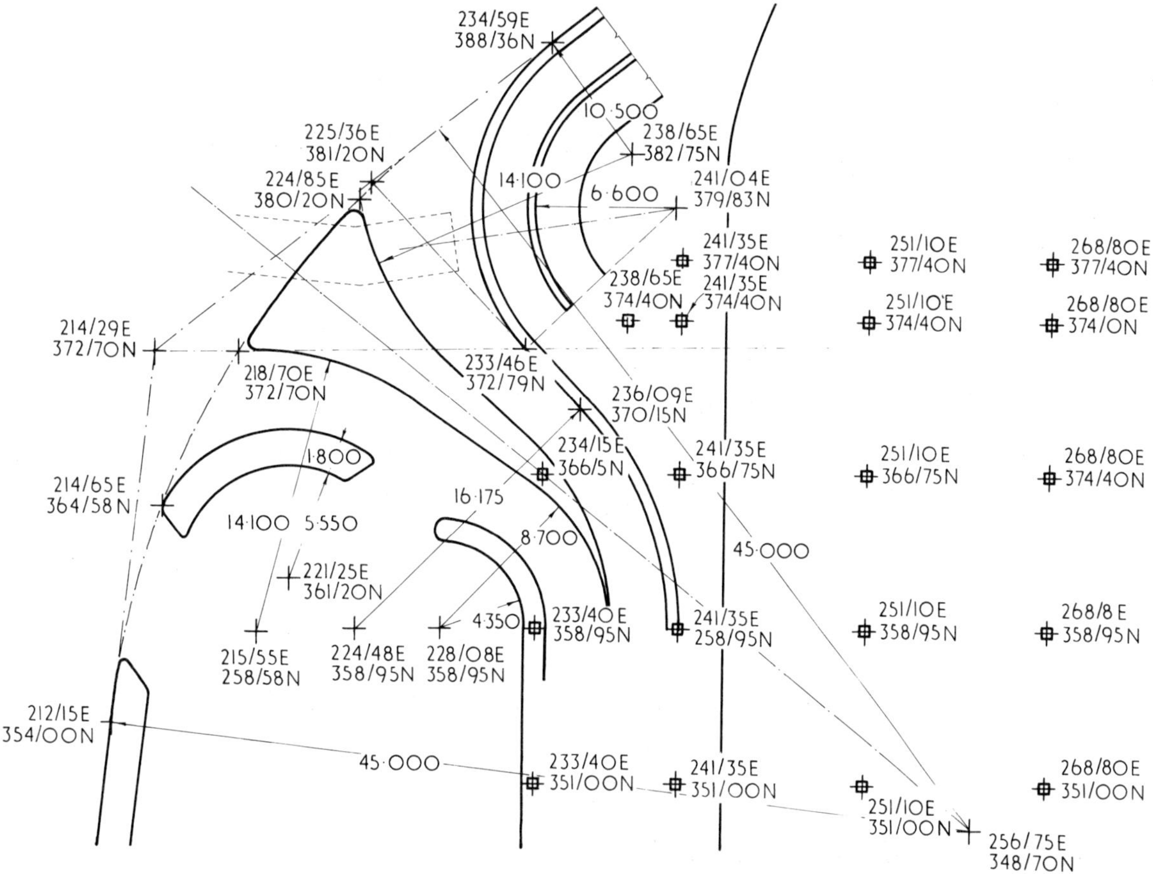

5 *Diagram is a reproduction of part of setting-out drawings prepared to accompany working drawings shown in 4. Calculated co-ordinates were used as considerable precision was necessary*

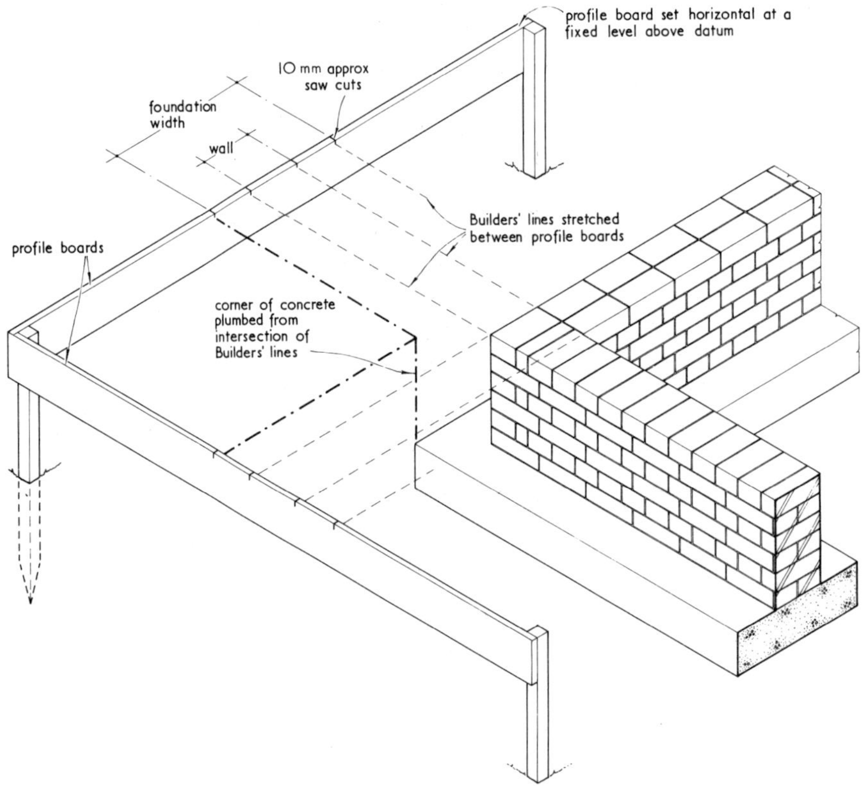

6 *Essential features of profile boards at the corners of buildings are shown. Single profile boards as shown in plan are used for cross-walls (continued)*

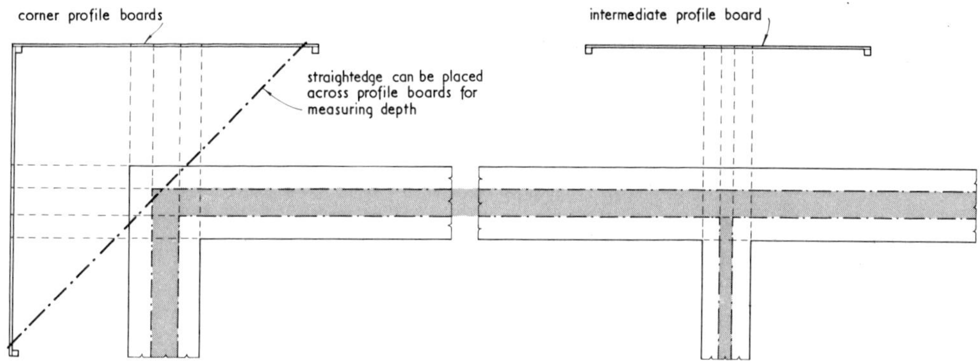
corner profile boards
intermediate profile board
straightedge can be placed
across profile boards for
measuring depth

plan

ordinates from fixed reference points such as existing buildings. The precise location of these points—within the permitted maximum error—is determined by driving wire nails or panel pins or incising a cross on the heads of the pegs, as in the case of the baseline. The number of key points established at any one time will depend upon the job in question since in some instances setting-out may have to be done in stages.

Profile boards Once the key points have been set-out, checked and agreed by the architect, their positions are usually fixed from without the building perimeter by means of lines and profile boards. The method is explained in **6**. When profile boards have been set up the key point pegs are removed so that work can proceed.

Curved buildings When buildings have curved portions in plan, or parts laid out to a radius, the profile board method is inappropriate. In such cases various expedients have to be used and contractors' methods tend to vary according to their expertise. Wide use is made of templates for curved work and sometimes centres for radii are marked with a peg. Some useful methods are given in **7**, **8** and **9**.

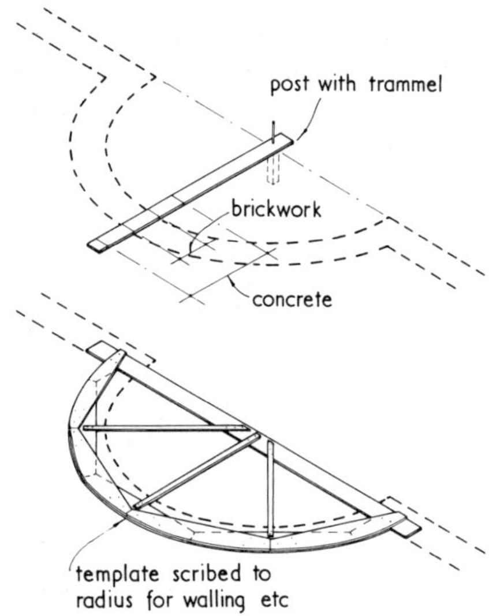

7 *Small semi-circular bays are set out either by means of a trammel or a template*

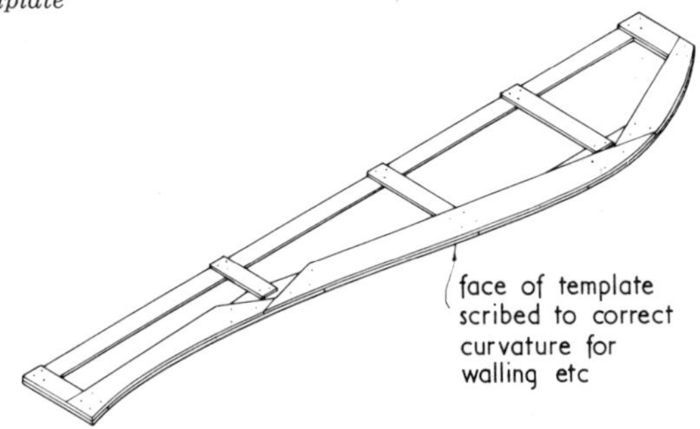

8 *Short irregular curves can be set out with suitable templates similar to that shown*

9(**i**) Large radius curves can be set-out from suitably placed pegs by means of a triangular template. This is basically an isosceles triangle whose base fits between pegs, whose apex marks a point on the curve (*i*) and thus when placed successively between a series of suitably located pegs (*ii*) traces out the required curve.

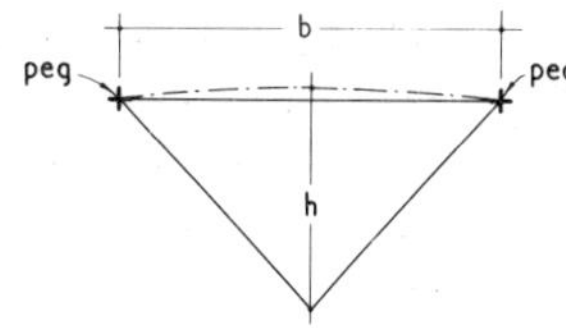

9 *Set of drawings and associated text captions:*
(i) *Basic triangle*

9(**ii**) The position of the pegs has to be calculated according to the size of template whose key dimensions will be b = base and h = perpendicular height **9**(**i**).

In architectural practice most curves are arcs of a circle of which the beginning and end are fixed dimensionally. The first step in preparing for setting-out is to find the angle w subtended at the centre from which the relevant arc was struck. If the location or co-ordinates of points A, B and C (*iii*) are known the angle w can easily be calculated.

In this example $\sin w = \frac{60}{120}$ and $w = 30°$.

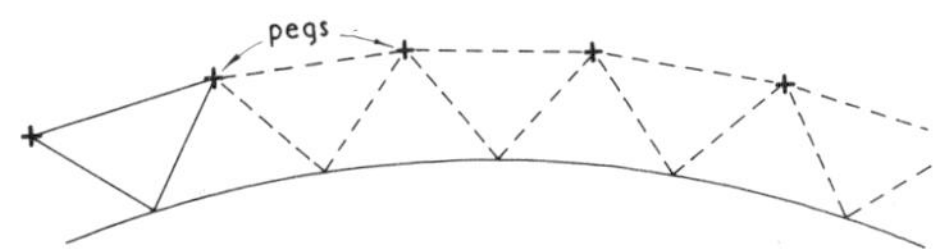

(ii) *Curve from repeated positions of basic triangle*

M

9(**iii**) The next job is to find the dimension *w* for the setting-out template. This is most convenient at about 3m. This means, in effect, that the setting out triangle's base dimensions *b* must be worked out to subtend an appropriate angle at C (the arc's centre point). In other words the angle ACB (in the example) must be subdivided so that *b* approximates to 3.

Since AB is roughly 60m then the angle ACB must be sub-divided into 20 (or 21) parts ie the angle subtended by *b* must be $30 \div 20 = 1\frac{1}{2}°$. Thus *b* (*iv*) becomes the base of an isosceles triangle of sides $r + h$ and apex angle $1\frac{1}{2}°$.

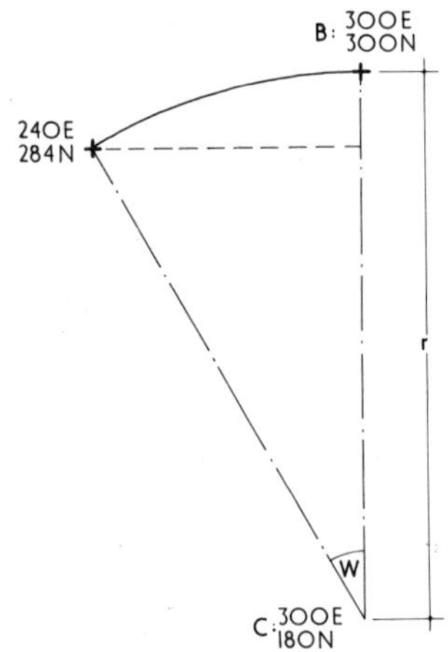

(iii) *Example of segmented curve*

9(**iv**) and **9**(**v**) If h is fixed (arbitrarily) at 2m, b can be calculated:

$$\frac{a}{\sin A} = \frac{b}{\sin B} \text{ or } \frac{b}{\sin 1°30'} = \frac{120 + 2}{\sin 89°15'}$$

and b = 3·194m

Given the dimensions b and h for the setting-out triangle, and the co-ordinates of the centre from which the arc is struck, co-ordinates for the setting-out pegs can easily be calculated by simple bearing and distance. In the example given it has been assumed that the line from C (v) to peg 1 runs parallel with the reference meridian ie grid north.

The calculation is:

peg 1 300 E 302N by inspection

peg 2 Easting: 122 sin 358°30′ = −3·194

Northing: 122 cos 358°30′ = −121·96

Co-ordinates: 296/81 E 301/96 N

peg 3 Easting: 122 sin 357° = −6·385

Northing: 122 cos 357° = 121·84

Co-ordinates: 293/61 E 301/84 N

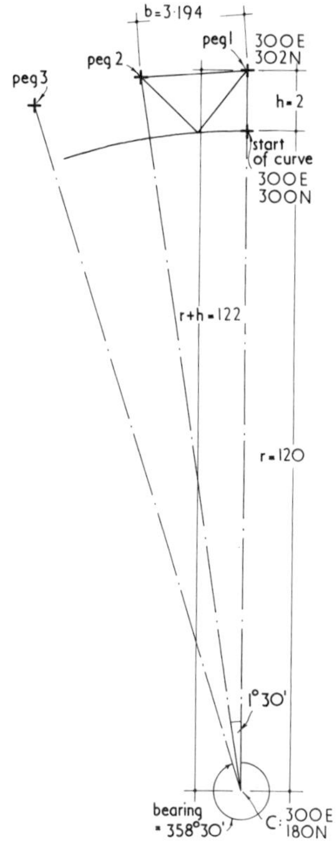

(iv) *To find size of basic triangle*

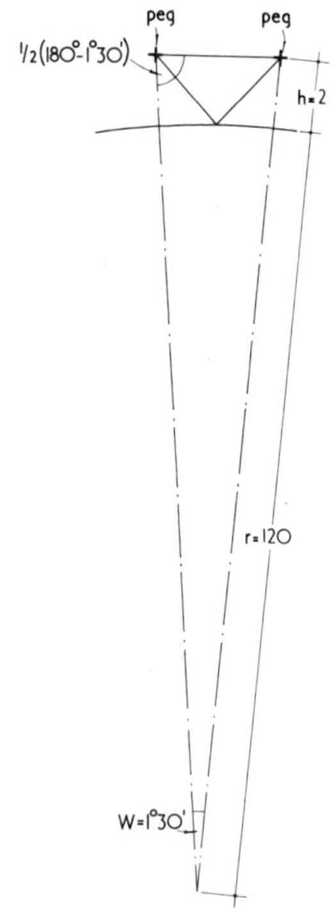

(v) *To calculate location of pegs for setting out*

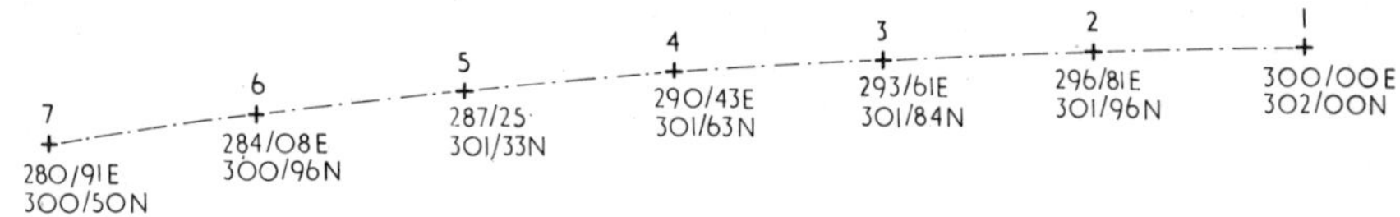

(vi) *Curve set out from calculated co-ordinates*

9(**vi**) The co-ordinates of subsequent pegs calculated in a similar fashion are:

peg 4 290/43 E 301/63 N

peg 5 287/25 E 301/33 N

peg 6 284/08 E 300/96 N

peg 7 280/91 E 300/50 N

The peg positions and curve plotted from these co-ordinates are shown in the diagram (*vi*).

9(**vii**) The triangular template can be set out with a central apex and two end apices as shown. Their outer sides will measure h but will be set to an inward angle $= \frac{1}{2}$ (180 − subdivided arc). In the example previously given this would be $\frac{1}{2}$ (180 − 1°30′) = 89°15′ and from a practical point of view a right angle would be satisfactory.

The perpendicular height of the basic triangle would not in fact be h but some smaller dimension d, as shown. The proper value for d is given by the formula:

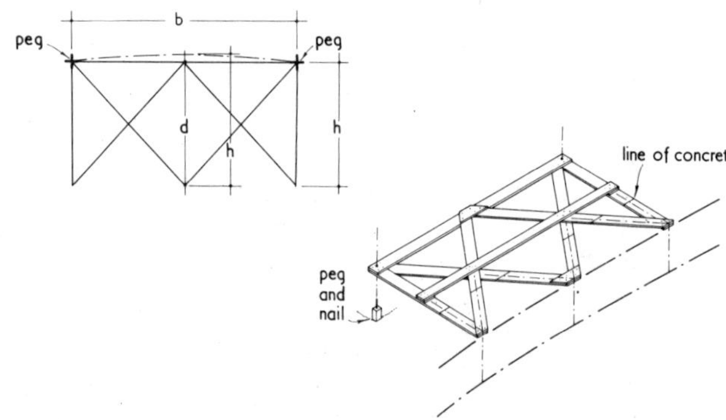

(vii) *Key dimensions of the actual template and as translated into practical template*

$d = (r + h) \cos \frac{1}{2}$ subdivided arc angle $-r$.
In the example previously given $h =$ and
$d = 122 \cos 45' - 120$
$= 1{\cdot}985$
Thus $d = h$ for most practical purposes.
This distinction should not be forgotten because it becomes more important as the radius of the curve r decreases.

It should be clear from these diagrams that it is important for the architect to give co-ordinates for the beginnings and ends of all curves, or fix them precisely by some other means. Curves other than simple small arcs of a circle up to about 30m radius (with an unobstructed centre) should be further defined by intermediate fixed points on the curve at intervals of from about 3m to 10m. These should be calculated rather than scaled from a drawing unless very precise draughtsmanship has been used.

Work below ground The setting out of foundations and walls below ground is done by plumbing (vertically orienting) the line of trenches and walls down from the lines stretched between profile boards or from templates etc. Normally a plumb-bob and line is used, but for deep work, or where high accuracy is called for, optical instruments may have to be used.

Setting out drains

Location and direction The points for which the most accuracy is needed are usually those at which soil drains enter and terminate in buildings, and the location of manholes. It is usually left to the contractor to work out the position of pipe outlets from other data provided by the architect. However the use of prefabricated heart units or connections for special pieces of equipment, as in hospitals, factories and other commercial and industrial premises, are likely to call for precise determination by the architect and a check on the contractor's setting-out.

Short lengths of trench can be excavated, as regards direction, by eye, but the following procedure is used for anything over a metre in length. The centre line of the drain is marked with a wooden peg at each point where it changes direction. This will usually be at manholes for which the location should be determined by the architect, either by co-ordinates, some kind of triangulation, or by reference to a fixed point or points on the site. Other changes of direction should be determined in the same way.

The sides of each trench are then pegged out at half their total width on each side of the centre line so that a cord stretched along the faces of the pegs will indicate where the spade or pick is to be inserted to mark the trench edges.

Drain levels Invert levels at each manhole, outlet, sewer connection or change of direction must be determined by the architect. The contractor normally works out the extra depth required between invert level and trench bottom and is required to see that the latter follows an even gradient from one point of given invert level to the next.

The method used to set out drain trench gradients to the correct level is illustrated in **10**. The levels transferred by boning rods and sight rails to the bottom of a trench are usually left 5 or 6cm above the required level until just before laying the drain. The architect or clerk of works should check the levels of sighting rails, the length of boning rods and accuracy of the trench bottom just before the work of laying pipes is begun. When manholes are complete invert levels should be checked directly from site datum.

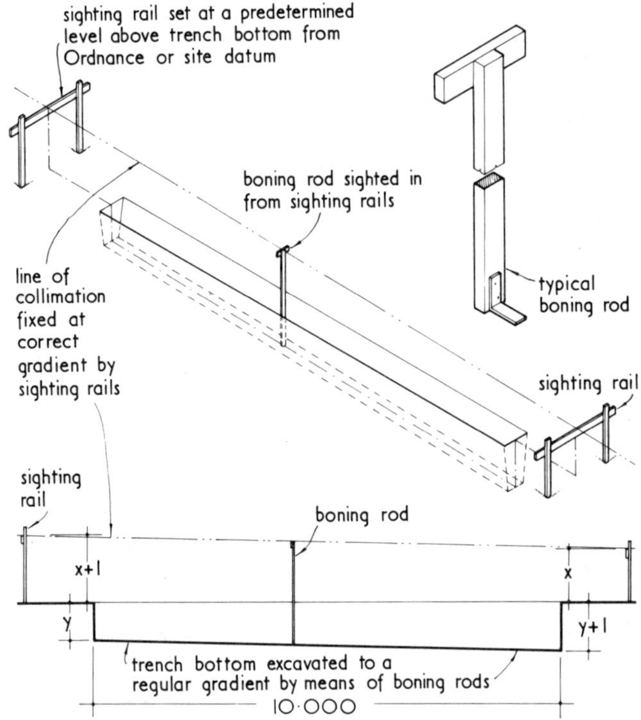

10 *Diagram illustrates general principle used for excavating trenches to regular gradient*

Foundation and basement levels

Profile boards The top edges of profile boards should

all be at the same level wherever possible. This not only provides a useful check on the horizontality of walls just above ground level but also simplifies setting-out of levels for excavation trenches with the aid of a boning rod or rods of standard length(s).

Procedure Pegs are driven at the corners in the bottom of foundation trenches after these have been excavated to the correct depths. The pegs are tapped down until their heads are at the correct level for the top of concrete strips or slabs. Various methods are used to fix this level but the commonest, which is quite adequate for most work is to measure down from a straight edge laid across the relevant pair of corner profile boards **6** (whose level relative to site datum must, of course, be known). Once the corner pegs have been adjusted for level, intermediate pegs are driven and their levels by reference to the corner pegs. This can be done either by using boning rods or by a straight edge and mechanics level. The pegs are left in until the concrete has been poured and levelled. It is important to see that they are withdrawn before the concrete has set or they may become responsible for the spread of dry rot, or rise of moisture even if dry rot is unlikely.

Checking levels The architect or clerk of works should check the level of the bottom of the excavations, preferably with an instrument, by reference to site datum, and again after concrete has been poured. This is preferable to checking peg top levels since the latter may have been accidentally knocked down a little in the course of the work.

Calculations for setting out levels

Profile boards The levels at which profile boards are set is mainly a matter of convenience, usually around half a metre above ground level on level ground or from half a metre upwards on sloping ground. One board should be related to site datum in the ordinary way with a level. The others can then be set so as to obtain the same staff reading on each.

Sighting rails Sighting rails have to be set at predetermined levels above invert level. This is done separately for each run of drain, ie each complete series of gradients and requires a certain amount of calculation once the first sighting rail has been put in position and its level found relevant to site datum. It saves much time and trouble to adopt a standard method. One which has been found satisfactory is related to the collimation method of levelling (see Chapter 12), and illustrated in **11**. Once the level of the first sighting rail has been found and those of the other sighting rails calcu-

Reading to site datum	Line of collimation	Required reading	Reduced levels	Remarks.	Sighting-rail levels.
Enter reading to site datum for each time level is set up.	Enter reading to site datum *plus* reduced level of site datum. The object is to find the height of the line of sight above datum.			Enter sufficient information to identify the point to which levels refer. This can be done in the office	Enter reduced levels of sighting-rails. i.e. reduced level *plus* length of boning rod.
			Enter level of site datum and then enter required levels taken from drawings. This is best done in the office.		
		Enter required levels subtracted from the line of collimation. i.e. finished levels or levels of sight-rails. — for drainage or other gradients.			
Example					
1·695	18·033		16·338	Site datum·Step of St. James St. baths.	
		2·973	15·060	Foundation (earth) level @ 357/236	
		1·743	14·490	Invert at M.H.1.	16·290
		2·163	13·920	Invert at M.H.2.	15·720
		2·943	13·290	Invert at M.H.3.	15·090
		1·350	16·683	G.F. level; top of screed, Entrance hall	

11 *Calculations for setting-out levels*

lated, setting levels is merely a matter of adjustment to the predetermined values.

Vertical orientation

Accurate vertical orientation The need for high accuracy in vertical orientation or plumbing arises whenever deep basements, high level cantilevers or overhangs, tall shafts and tall buildings are being constructed. Precision is important not only for structural efficiency and the high costs of mistakes but also because human safety may be involved both during and after construction. The traditional plumb-line and conventional built-in optical plummet are inadequate in such cases. Special attachments for instruments such as theodolites also have a limited precision and so, for major constructional works, special optical plummets designed specifically for such purposes are best used (Chapter 2.1). Sometimes very high grade theodolites are used, with the telescope pointing vertically (upwards) and an additional special eyepiece fitted to make it possible to see through the telescope in that position.

Optical plummets These may be used to sight either upwards or downwards or both according to the instrument available. When sighting upwards the instrument is set over the relevant reference mark and a target which is shifted until its centre coincides with the line of sight. When optical plummets are used to sight downwards, or both downwards and upwards from an intermediate level, the procedure differs somewhat because the instrument must be within the perimeter of the building and thus away from ordinary site reference points. The problem is met by using 'insets' from the main setting-out line (see next para).

Insets In the construction of tall buildings a system of reference lines parallel to but inset from the main setting-out lines must be established at upper floor levels. This is usually done by leaving small sighting holes about 150mm square through each floor and close to each corner of the building: they are best placed about 600mm in (in each direction) from the setting-out lines. These inset lines are fixed at ground level in the ordinary way from profile boards about 0·8m in from the main setting-out lines **12**.

Inset observations When an instrument capable of observation both up and down is used the best results are obtained by setting up the instrument at the halfway level. Ten floors is usually considered the limit for one set of observations and so, for ten floors, the optical plummet would be set up on the fifth floor **13**. When sighting downwards a staff, rod

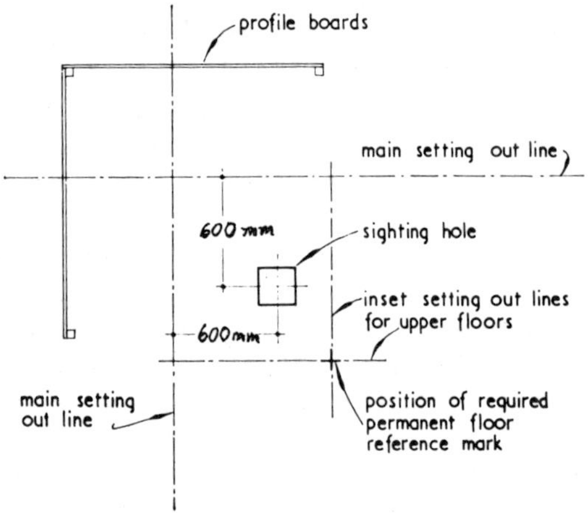

12 *Inset lines and sighting holes for setting-out in tall buildings*

or scale is placed at ground level so that it can be seen through the sighting holes from above. It is placed at right angles to and with its zero mark on one of the inset setting-out lines **14**. The scale is observed and read twice: the second time with the instrument reversed through 180 degrees, eg if the two readings were 0·758 and 0·764, the mean reading of 0·761 would be recorded. The procedure is then repeated with the staff or scale moved through 90deg to obtain the location of the inset line at right angles to the first.

The whole procedure given above is then repeated for each corner of the building, after which reference points on the first, second, third, fourth and fifth floors (downward sighting) are fixed thus:

An assistant places the staff (or rod etc) over the first floor sighting hole approximately at right angles to one of the inset lines—right-angularity can then be corrected by reference to the instrument reticle cross-hairs. The staff is moved, thereafter, so that the reading corresponding with the initial ground-level position intersects the centre of the reticle. The point on the floor where zero comes is then marked, the assistant does the same thing a second time with the instrument reversed through 180deg and eventually marks the mean reading: the procedure is then repeated with the staff moved through 90deg, and so on for each corner of the building.

The same procedure is followed when sighting upwards for the sixth, seventh, eighth, ninth and tenth floors, but there is the complication that the staff is now face downwards. This makes it necessary either to use a transparent scale or to have the

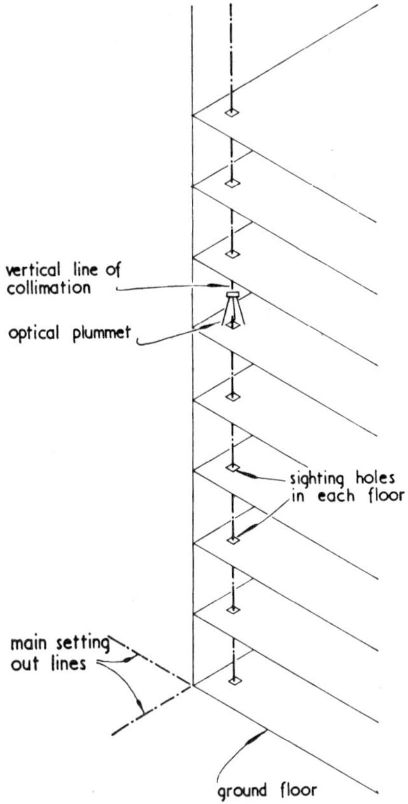

13 *Diagram illustrates use of optical plummet in tall buildings for vertical orientation through sighting holes*

zero mark extended round the staff or rod so that its position can be seen by the assistant handling it.

Inset lines on upper floors The location of upper floor inset lines can be fixed either by stretching the building line across the points previously marked **15** or by setting up a theodolite over at least half the marked points and directing an assistant with a pencil to mark the intersection of the approximate position of the inset lines and confirming it with a second sighting from the opposite corner. When located, the intersections of inset lines should be checked by measuring all diagonals which should correspond very closely with those of the ground floor.

Limits of error The accuracy requirements imposed by local authorities, consulting engineers and specialists will vary in different parts of the country and with different jobs: they should be incorporated into specifications or bills of quantities. The crucial requirements will relate to the greatest permissible displacement of the main supporting stanchions or columns from verticality at the top of a building. Up to twenty floors the outside limit is likely to be less than 25mm. Above twenty floors a proportionate increase of tolerance, up to a limit of 50mm is correspondingly likely to be expected.

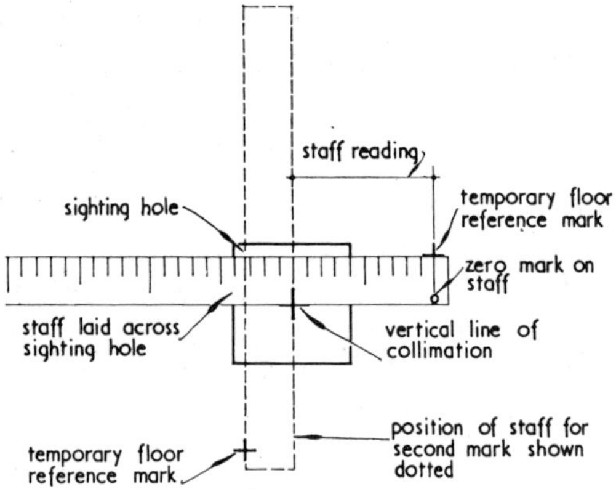

14 *Measuring rod or staff placed over a sighting hole to locate an inset line*

In order to keep within these limits considerably better accuracy is needed for the location of setting-out lines because small errors in measurement cannot normally be avoided and the total error multiplies with each operation. Thus 2mm at the tenth floor should normally be regarded as the maximum permissible dimensional error for the inset setting-out lines relative to the corresponding ground floor lines at any one point.

Sources of error Common sources of error may be summarised thus:

1 failure to set up the instrument correctly;
2 mistakes in measurement when establishing the inset setting-out lines;
3 mistakes in reading the staff at upper floor levels;
4 inaccuracy in marking reference points on the floor;

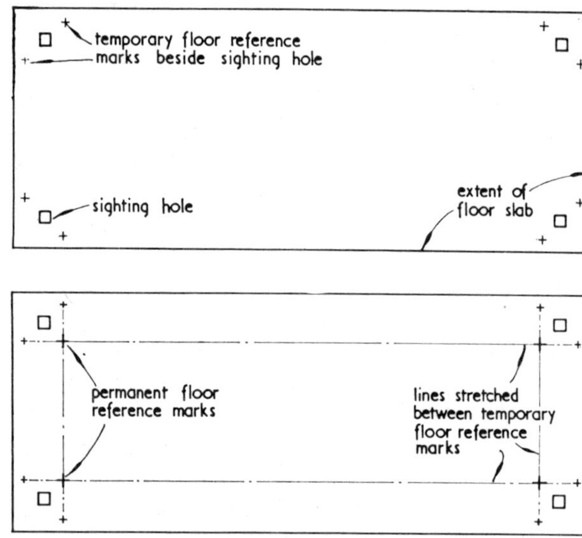

15 Diagrams illustrate inset lines located from marks fixed by readings through sighting holes.

5 inaccuracy in laying out the inset setting-out lines from the key points.

Internal walls and partitions The inset lines at each floor level can also be used for setting out those non-structural walls and partitions which cannot easily be related to structural features in the normal way.

For structures based on the principle of dimensional co-ordination and using prefabricated components related to the dimensional framework, accuracy is even more important, and accuracy in setting-out must be insisted upon.

Setting out must obviously be done with a steel tape, possibly with the BRS tension corrected handle; see chapter 8; and based on intersecting grids. Readings between points on the grid should be made by means of running dimensions marked alongside the static tape. To obtain a further check the zero point of the tape should be reversed and readings marked off again with the tape running in the other direction. The mean values of the two procedures will reduce the risk of errors to a minimum.

The use of a theodolite will ensure that the building is square—both in horizontal and vertical planes **20**. It is equally important to insist on frequent control of the gridded structure by checking

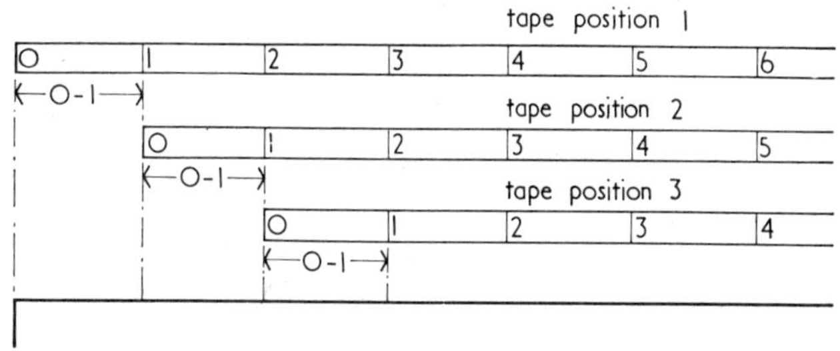

16 *Method of double checking dimensions by moving tape along one unit per time and noting any discrepancies. Tape may also be reversed to measure from opposite direction*

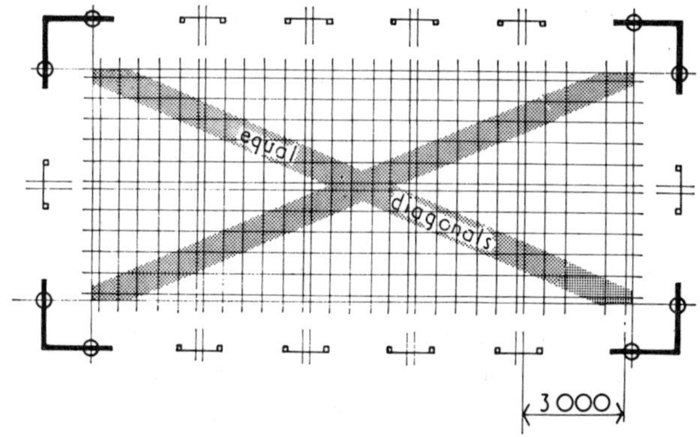

17 *Setting out for schemes using dimensionally co-ordinated framework and preformed components closely related to this framework, may impose rules of procedure for checking, which being part of accepted routine could be regarded as unnecessary. It is, however, surprising how frequently exhortation on use of theodolite and checking of diagonals is ignored on site*

diagonals **17**.

The use of the planning and structural grid in a dimensionally co-ordinated scheme is a logical follow-up from the design stage. The employment of grid lines related to a reference system will enable the setting out to be more accurate and will help to locate the critical terminal points of service outlets **21**, **22**.

Paragraph 4.02 from the BSI document PD 6440 Draft for development—accuracy in building, shown in table I, is perhaps the best part of that rather sad document, with Appendix C 'Use of grids in construction' and Appendix D 'Some typical setting out methods'.

It is worth quoting this table if only to prove that the more primitive the instrumentation, the more primitive the results.

18, 19 *Sites like drawings become dimensional matrixes within which framework components are located and fixed. Most important activity—setting—so frequently placed in hands of simple operatives must be brought into high importance category of site management. Dimensionally co-ordinated grid references of designer must be expressed in setting out stage. Use of standard templates and jiggs to aid accuracy of setting out must be encouraged. Here, timber template for holding-down bolts set out in relation to grid system helps overall efficiency*

Table 1 Acceptable deviations in setting out (from PD 6440)

Instrument	**Permitted deviation**	**Comments**
Linear dimension		
30m carbon steel tape	±5mm up to and including 5m; ±10mm over 5m and up to and including 30m and over 30m pro rata	With sag eliminated and slope correction applied and extremes of temperature avoided
Angular measurements		
Optical square	±15ft (±65mm in 15m)	Not suitable for accurate setting out
30m carbon steel tape	±2ft (±10mm in 15m)	Normally in right angles only. With sag eliminated and slope correction applied
Tripod mounted optical square	±1ft (±10mm in 30m)	
Vernier theodolite calibrated to 20in	±40in (±10mm in 50m)	Mean of two sights, one on each face
Microptic theodolite (with optical plummet) calibrated 20in	±20in (±5mm in 50m)	Scale readings estimated to nearest 5in. Mean of two sights, one on each face
Ditto, calibrated to 1in	±5in (±2mm in 80m)	Mean of two sights, one on each face
Verticality		
Spirit level	±10mm in 5m	
Plumb bob (3kg)	±5mm in 5m	If used in still conditions
Theodolite (with optical plummet) and diagonal eyepiece	±5mm in 50m	
Optical plumbing device	±5mm in 100m	This is a special automatic instrument
Vertical dimensions (levels)		
Spirit level	±5mm in 5m distance	
Water level	±5mm in 15m distance	
Light self-levelling level	±5mm in 25m distance	
Optical level	±5mm per single sight of up to 60m and ±25mm per km	Suitable for most site setting out
Automatic level	±5mm per single sight and ±10mm per km	Not suitable where vibration is present (eg tall buildings)
Precise level	±2mm per single sight and ±10mm per km	

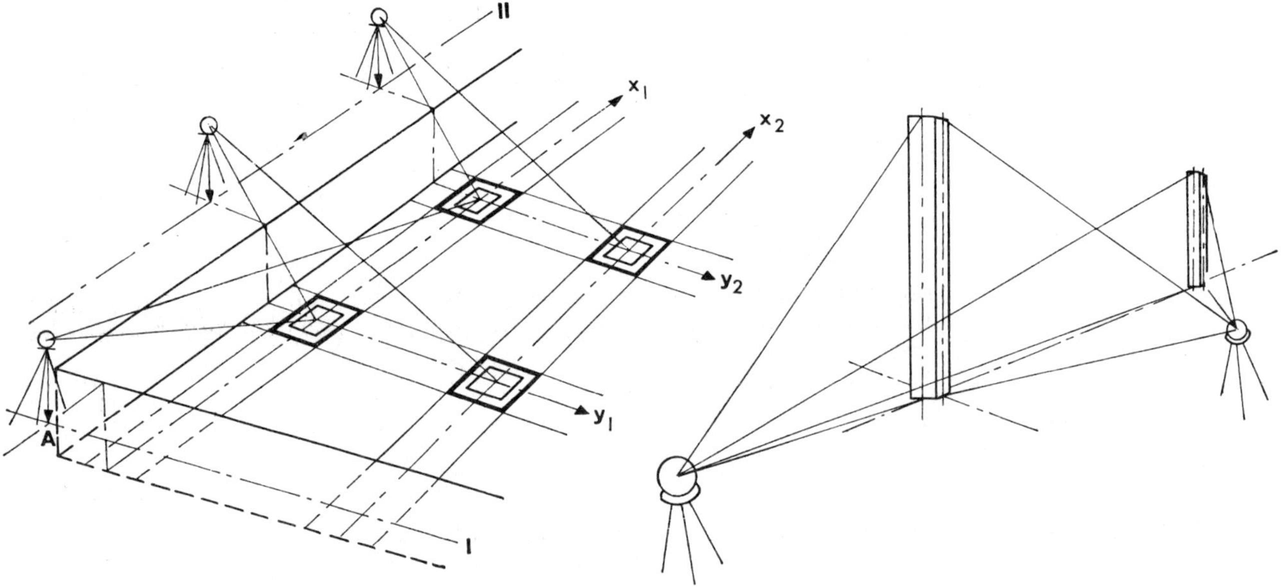

20a, b *Quality control begins with setting-out sequence—involving right type of instrumentation and correct procedure. These examples show establishment of a horizontal and b vertical datum lines as well as structural grid lines and zones by means of theodolite*

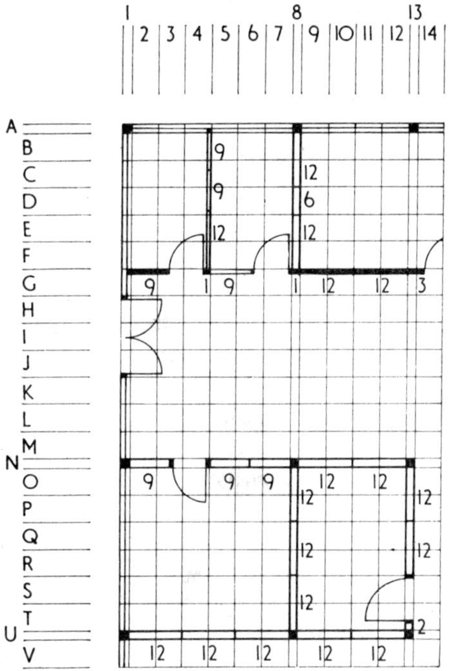
1 2 3 4 5 6 7 8 9 10 11 12 13 14
A B C D E F G H I J K L M N O P Q R S T U V
9
9
12
12
6
12
9 1 9 1 12 12 3
9 9 9 12 12
12 12
12 12
12
2
12 12 12 12 12

21

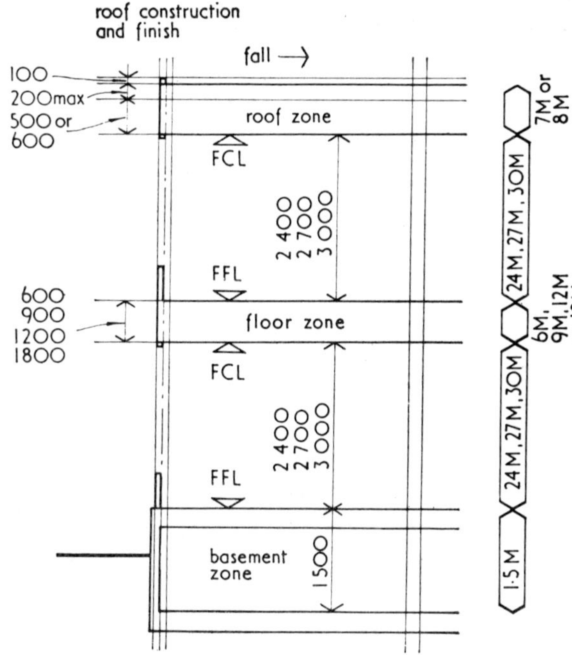
roof construction and finish
fall →
100
200max
500 or 600
roof zone
FCL
2 400
2 700
3 000
FFL
600
900
1200
1800
floor zone
FCL
2 400
2 700
3 000
FFL
basement zone
1500
7M or 8M
24M, 27M, 30M
6M, 9M, 12M or 18M
24M, 27M, 30M
1·5M

22

N

Chapter 15
Surveying techniques: Services available to the surveyor

This chapter gives details of the services offered by Ordnance Survey apart from the production of maps and plans. These services arise from the survey's normal work of triangulation, levelling, aerial and field surveys, drawing, printing and publication. The services and their costs are liable to change. Except where otherwise mentioned, requests should be made to The Director General, Ordnance Survey, Romsey Road, Maybush, Southampton SO9 4DH. The chapter includes sections on copyright and publications, and lists the addresses of the region offices.

ORDNANCE SURVEY SERVICES

Note that horizontal control consists of triangulation stations; minor control points are known as permanent traverse stations and revision points. The technique of surveying by triangulation and terms used in surveying are explained in Chapter 11.

Lists of national grid retriangulation co-ordinates

Are compiled by national grid 10km squares and give the description (eg pillar, buried block) of each permanent traverse station in that square. *Cost* £1 a page of which, depending on the density of the triangulation, there may be up to four for a 10km square.

Triangulation station descriptions

Give the details necessary to locate the point, and information on the ownership of the land where the station is situated. (Permission should be obtained from landowners or tenants before using a station on private land.) **1**. If a purchaser finds that a station is unusable he may obtain details of an alternative point without further charge.
Cost £11 per triangulation point.

Minor Control Information

For a permanent traverse station the information availabe includes

(i) The number of the station
(ii) The type of mark
(iii) A written and graphic description grid bearings to local reference objects
(vi) The National Grid co-ordinates (eastings and northings) to 2 decimal places of a metre

For a revision point the information available includes

(i) The number of the point

Ordnance Survey

Romsey Road Maybush Southampton SO9 4DH

Telephone 0703 (Southampton) 75555 ext 518/519

Your reference

Our reference

Date

TRIANGULATION STATION INFORMATION

Name of Station Steynton Ch Twr — K.M. Number SM9107

Station No SN/02/T72 — 1-Inch Sht. No 138/151 (7th Series)

Type of Mark Bolt — Fl. Br. No. — County Pembs

Order in Trig. Network 3rd — Date of Co-ordinates 1966

Nature of Ground/Roof Lead & Tiles

Route and Access See Vicar

Name, Addresses of Owner, Agent and Tenant:-

Owner

N.G. Ref

Agent The Vicar
Steynton Vicarage
Milford Haven

N.G. Ref SM918078

Tenant

N.G. Ref

General Aspect & Obstructions

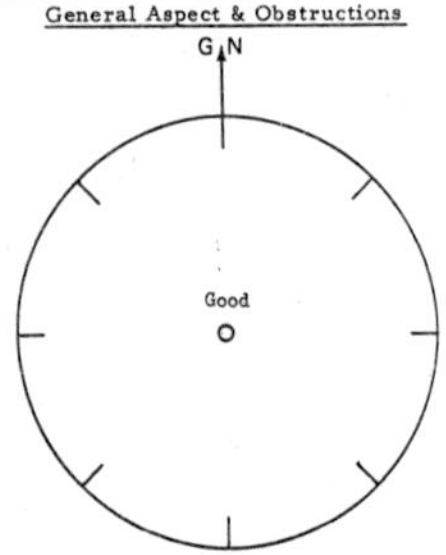

General Information

Name of Station:- Steynton Ch Twr — Station No. SN/02/T72

DESCRIPTION

(Measurements in metres).

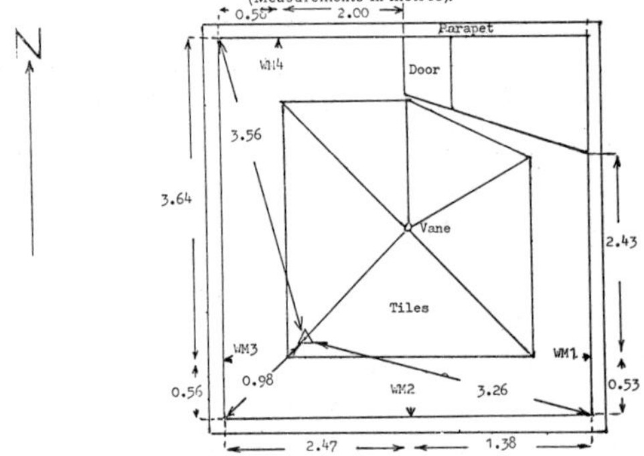

National Grid Bearings to:-	Km. Sq.	Dist. Approx. Kms.	°	'	"
Fort Scoveston (Pillar)	SM9406	3.0	114	42	19
Whetstone Hill Ch Spire	SM9800	9.7	138	46	11
Milford Haven Ch Twr (FS)	SM9005	2.3	204	10	02
North Neeston (Pillar)	SM8807	3.4	269	10	51
Bolton Hill (Pillar)	SM9111	3.4	358	33	52

Height of Building (G.L. to Station Mark) No Ht m Height of Parapet m

Witness Marks Measurements (metres)

	Direct	Horizontal
Bolt to W.M. No. 1		3.198
" " " " 2		1.942
" " " " 3		0.714
" " " " 4		3.461

	Direct
W.M. No. 1 to No. 2	1.552
" " 2 " " 3	2.551
" " 3 " " 4	3.654
" " 4 " " 1	Not Possible

National Grid Re-Triangulation Co-ordinates

Bolt E 191 767.86 — N 207 816.59
E — N
E — N
E — N
E — N

Altitude on Newlyn Datum to Fl. Br. or Top of Pillar Ft m

The information on this Triangulation Station with the exception of the Co-ordinates, Altitude and Grid Bearings was completely revised January 1966

1 *The face and reverse of a typical triangulation station description*

(ii) The type of mark

(iii) A written and graphic or photographic description

(iv) The National Grid co-ordinates (eastings and northings) to 2 decimal places of a metre.

For points not indicated on the published maps, the information normally available includes

(i) The number of the point

(ii) A graphic description

(iii) Angles referred from an adjacent station of similar type with distances to local objects such as corners of houses, etc.

(iv) The National Grid co-ordinates (eastings and northings) to 2 decimal places of a metre.

As a result of building development or other causes, control points may be destroyed or damaged. Information that has been checked is known as 'Verified'. Where 'Unverified' information, which is cheaper, is supplied the Ordnance Survey can neither give a guarantee of the existence of a point nor make a refund of the charge.

Cost Permanent traverse station field verified £5·50
Permanent traverse station unverified £2·20
Revision point field verified £5·50
Revision point unverified £1·65
Other minor control point, unverified only £2

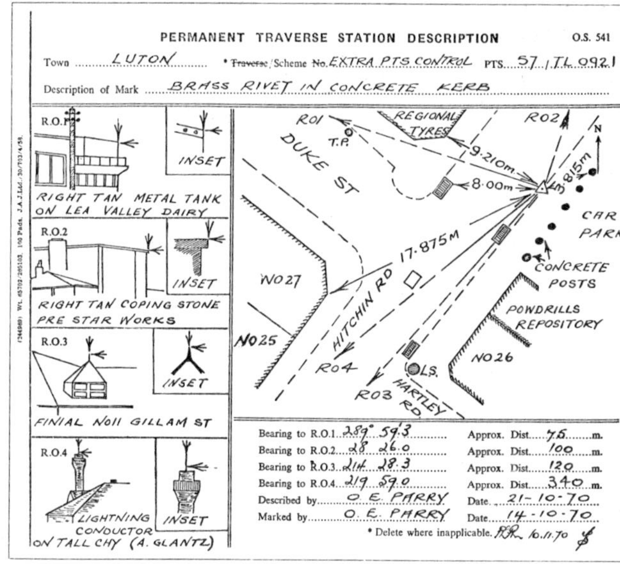

2 *A typical permanent traverse station description*

Bench mark lists

Describe briefly the bench marks in 1 sq km on the national grid with their national grid reference, altitude to 0·01ft and 0·01m referred to the Newlyn datum, height above ground level and date of levelling. The positions and values of bench marks

are shown on the 1/1250 and 1/2500 scale national grid plans, but the bench mark lists are more up to date as new editions of plans are not published whenever a bench mark is destroyed or new levelling information becomes available **3**.
Cost £1 a page.

National grid co-ordinates for county series plans
Relate the county series plans to the national grid. The Ordnance Survey is recasting on national grid sheet lines, and then revising the 1/2500 scale county series plans. Only 30 000 of the total 170 000 sq km plans have been completed and the county series plans will be in use for some years yet. National grid co-ordinates are available of the sheet corners of any county series 1/2500 scale plan or six-inch sheet.
Cost £1·10 a plan or sheet.

Aerial photographs
Can be supplied only to government departments, local authorities and other official organisations when the request cannot be met by the Department of the Environment or the Scottish Development Department. Libraries of aerial photography are maintained by DOE and SDD, to which requests for information and prints should be addressed.
The Ordnance Survey has considerable experience of aerial survey and invites consultation from local authorities who are thinking of placing contracts for aerial photography or survey.

Reduction Service
Reductions of recently revised 1:1250 plans are available to be used in up-dating 1: 2500 plans. They are supplied on film or paper to suit the type of reproduction equipment used by the purchaser.
Cost Paper copies £1·40
Film copies £5
Master grid for use with film copies £3.50

Advance revision information
Is available in the form of surveyors' up to date field sheets resulting from the continuous revision of the large scale national grid plans. Information can be provided only where national grid plans have been published.

Cost Prices at the scale of the field sheets are £6 on paper and £8 on film
Enlargements are £15 per half plan.

ORDNANCE SURVEY BENCH MARK LIST

BATCH 623 OS 609

BATCH 623 ALL BENCH MARKS ON THIS LIST FALL WITHIN KM SQ SW8150 DATUM NEWLYN PAGE 1

DESCRIPTION OF BENCH MARK	NATIONAL GRID TEN METRE REFERENCE	ALTITUDE		HEIGHT OF BM ABOVE GROUND		DATE OF LEVELLING
		FEET	METRES	FEET	METRES	
SW ANG BLDG BOSWELLICK E SIDE TRK	8104 5064	248.81	75.84	1.3	0.4	1953
GTP 11FT S HEDGE JUNC N SIDE LANE	8118 5036	242.11	73.80	1.6	0.5	1953
NE FACE 6FT E ANG BLDG SE SIDE LANE	8142 5025	166.50	50.75	0.7	0.2	1953
RIVET FOOTBR 3FT NE STR SE SIDE LANE	8154 5033	157.85	48.11	0.1	0.0	1953
GTP 32FT S HEDGE JUNC E SIDE LANE	8177 5027	292.04	89.01	1.2	0.4	1953
GTP HEDGE JUNC E SIDE TK	8179 5049	273.62	83.40	0.9	0.3	1953
S FACE 9FT SW ANG COTT NE SIDE LANE JUNC	8180 5005	230.88	70.37	2.5	0.8	1953

Date of issue 09/11/71

continued on page

In England and Wales, BM's levelled before March 1956 are based on the second geodetic levelling, and those levelled afterwards on the third geodetic levelling. Differences between altitudes derived from these two levellings may be found due to the differing adjustments, but for most practical purposes these are insignificant. An estimated correction to relate one levelling to the other, by areas, is available from the Director General, Ordnance Survey.

In Scotland BM's are based on the second geodetic levelling, with additional heights derived from the third geodetic levelling in areas not previously levelled.

To convert metres to feet, divide the metric value by 0·3048.

3 *Bench mark list*

Enlargements of large scale plans

These enlargements are available on clear or matt Kodagraph only as 1/500 scale enlargements of 1/1250 scale plans and 1/1250 scale enlargements of 1/2500 scale plans. Each enlargement is produced in two parts with a small overlap. Orders for part maps are accepted.

Requests should be made to Southampton or Ordnance Survey region offices. Commercial copying firms provide a similar service, usually at a lower price.

Cost Published sheets £8 per enlargement
Field documents £15 per enlargement

Supply of Transparencies

Many users of Ordnance Survey maps wish to keep a record of special information, such as the alignment of water mains, electricity or telephone cables, and this can best be done by recording the information on transparent copies of Ordnance Survey maps. Paper copies containing the latest information can then be produced provided reproductions conform to the regulations governing the reproduction of Ordnance Survey maps.

To meet this need the Ordnance Survey supplied monochrome copies of most maps on a transparent material. Transparencies of field documents of certain large scale maps can also be supplied.

Transparencies of large scale maps show all published detail except that contours are not shown in the case of the 1: 10 000 and 1: 10 560 (six-inch) scale maps. Similarly, transparencies of the 1:25 000 and smaller scale maps are made from outline editions and do not show contours.

Cost

Maps on National Grid sheet lines
1: 1250 sheet £4
1: 2500 1km × 1km sheet £4
1: 2500 2km × 1km sheet £5·50
1: 10 000 sheet £4·40
1: 10 560 sheet £4·40
1: 25 000 10km × 10km sheet £4
1: 25 000 20km × 10km sheet £5·50

Maps on County sheet lines
1: 2500 sheet £5
1: 10 560 full sheet £5
1: 10 560 quarter sheet £3

Other maps
1: 63 360 and smaller scales £6

Microfilms
Copies of 1: 1250 and 1: 2500 N.G. plans in the form of 35mm diazo negative microfilms mounted in aperture cards £1·20

ORDNANCE SURVEY COPYRIGHT

The need to reproduce Ordnance Survey publications in whole or in part is provided for by a scale of royalty charges and a series of licences designed to meet the needs of various authorities, organisations and the professions. No licence permits the holder to make facsimile copies of complete and unamended maps or plans.
The form of licence issued to architects (os 484) authorises the holder to reproduce Ordnance Survey maps by hand tracing only; or by hand tracing and mechanical means. Reproductions, which shall not exceed 500 sq in in size, are not permitted where Ordnance Survey published sheets would be suitable. The number of copies of any one extract is limited to twenty. Full information on copyright is given in leaflet os 23 Copyright Regulations, available from HMSO or os regional offices.

ORDNANCE SURVEY PUBLICATIONS

Publication report
Issued each month, listing the maps, plans and bench mark lists published during the previous month. The report is free but postage is charged.

Leaflets for Ordnance Survey services

No.	*Title*
1	Price List of Special Services
2	Triangulation and Minor Control Point Information
3	Levelling
6	Reductions of 1: 1250 Plans to 1: 2500 Scale
7	Notes on A.R.I. Maps
11	A.R. Information
12	Reductions of 1: 10 000 to 1: 10 560 and Enlargements of 1: 10 560 to 1: 10 000
13	Maps for Educational Purposes
14	Transparencies
15	Enlargements
16	Map Mounting Services
19	List of Agents
20	Metrication
21	Price List of Maps
22	List of Licensed Printers
23	Copyright Regulations

27 Ordnance Survey Services
33 35mm Microfilms of Large Scale Maps

ORDNANCE SURVEY ADDRESSES

Table I Addresses of Principal Ordnance Survey offices

Address	Telephone number	Grid reference
Headquarter's Office		
Ordnance Survey		
Romsey Road, Maybush		
Southampton SO9 4DH	Southampton 75555	SU 387148

Regional Organisation

For surveys operations the country is divided into six regions, with offices at the addresses shown below.

Address	Telephone number	Grid reference
South East		
Ordnance Survey		
Tolworth Tower		
Surbiton, Surrey	01 399 5191	TQ 197658
South West		
Ordnance Survey		
Block 1		
Government Buildings		
Flowers Hill		
Brislington		
Bristol BS4 5JZ	Bristol 78311	ST 623701
East Midland		
Ordnance Survey		
Block 7		
Government Buildings		
Chalfont Drive		
Western Boulevard		
Nottingham NG8 3SL	Nottingham 291111	SK 539411
West Midland		
Ordnance Survey		
Crown House, Bull Ring		
Kidderminster		
Worcestershire DY10 2EB	Kidderminster 65151 Ext. 330 or 274	SO 831767

North
Ordnance Survey
Wing A, Block 2
Government Buildings
St George's Road

Harrogate	Harrogate 4621	SE 300538

Scotland
Ordnance Survey
43 Rose Street

Edinburgh EH2 2NL	031 225 4522	NT 252739

GEOLOGICAL SURVEY AND MUSEUM

Publications and information

The main work of the survey is the production of the official geological map of this country. Maps are published on a scale of 1in to 1 mile with special areas available on a 6in to 1 mile scale. Photocopies of unpublished 6in maps may be obtained. A list showing availability of maps may be obtained from the museum (address below). Memoirs describing the geology, including details of sections, bore holes, and other data have been published for many 1in sheets.

Eighteen handbooks describe British regional geology; other memoirs deal with economic mineral resources including ironstone, metalliferous ores, feldspar, limestone, granites and coal.

Detailed information on all aspects of geology has been gathered from many sources and is freely available for all who wish to use it. As well as a comprehensive knowledge of the geological structure of the country, there is a collection of building stones from most quarries now working stone in Great Britain; specimens from many British sources now disused, which can be an aid to matching stone for building repair; and marbles and other ornamental stones from most countries of the world.

Inquiries should be made at or addressed to the Geological Museum, Exhibition Road, South Kensington, London sw7. The northern England branch is at Geological Survey office, King Road, Halton, Leeds, 15. The Scottish office of the Survey is at Southpark, 19 Grange Terrace, Edinburgh 9.

METEOROLOGICAL OFFICE

Publications

For publications useful to architects see 'Guide to site investigation' sfB (1): Aa9 para 35 Selective bibliography (AJ 6.10.65).

Reports and other services

In addition, the Meteorological Office publishes the *Monthly weather report* which gives a summary of observations compiled from returns of official stations and voluntary observers. Long period averages and frequencies can be worked out from a series of the reports for places in the report. The *Annual summary of the monthly weather report* contains data and information for the year as a whole, largely similar to that in the monthly reports, but with some additional tables.

Information helpful for design purposes is available on request. Examples are: extreme wind speeds likely to occur once in a stated number of years, for design of structures; frequency tables of wind directions (and associated speeds); frequency table of dry bulb and associated wet-bulb temperatures, for designing ventilating and air-conditioning plant.

A charge is made for answering inquiries, depending on the time spent on the work. Usually the charges are small, but estimates of the cost can be given in advance.

All written inquiries should be addressed to the Director General, Meteorological Office, London Road, Bracknell, Berks, unless relating to Scotland or Northern Ireland, when they should be addressed respectively to the Superintendent, Meteorological Office, 26 Palmerston Place, Edinburgh 12, or to the Senior Meteorological Officer, Tyrone House, Ormeau Avenue, Belfast. Visitors can extract data free of charge at these offices. A limited amount of data is also available at the following Weather Centres: 284–286 High Holborn, London wc1; 160 High Street-below-Bar, Southampton; 56 Royal Exchange, Manchester 2, and 118 Waterloo Street, Glasgow c2.

Part 3
Site investigation

Chapter 16
Guide to site investigation

This chapter provides a check list (in column 1) and advice, information and references (in column 2) covering, first the broadest principles involved in assessing the suitability of a site for a project; second, the more detailed considerations of the site and of its climate and soil structure. The items included in the guide will be relevant to most types of project. The chapter is divided into two main sections:

1 Preliminary site information:
to establish that the site is suitable for the project
2 Detailed site information:
to gather full information about the site before the project design is undertaken

These two sections are essentially parts of one process and in some site investigations (eg where feasibility is already established) the division may be ignored and the guide used as one continuous procedure.

Bibliographic references in col 2 of this guide are given a shortened title for identification only. The full title, with details of publication, is given in the Bibliography at the end of the book.

1 Preliminary site information

Column 1 *Main guide headings*	*Column 2* *Advice, information, references*
1 Names and addresses	Record names, addresses and telephone numbers of individuals and organisations listed in col 1. Others will be required at a later stage and are mentioned in para 10 and 11 in the second section of this guide. (*Kelly's directory* [Aa2] may prove helpful)
Site address	Including a map reference if possible (see also para 9 for list of available maps and plans)
Client	Establish who, in the client's organisation, is responsible for the project or will act as liaison between architect and client
Agent	Person acting for the client in purchasing the site
Vendor and/or tenant	Permission may be required from this person to gain access. Site or buildings may or may not be occupied
Solicitors	Both client's and vendor's solicitors or other persons concerned with the drawing up of lease or freehold documents

Architect	If previously employed
Authorities: local: authority engineer and surveyor planning officer county: planning department Department of Trade and Industry regional office conservancy: river board National Trust, canal authority etc	Matters to be discussed with these authorities are listed in para 4. More detailed local authority inquiries are listed in para 11
Undertakings: gas board electricity board water board GPO telephones transport police parkways	Matters to be discussed with these undertakings are listed in para 5 (preliminary information) and para 12 (more detailed information)

2 General inquiries

Other architects	As a matter of courtesy it should be established in writing

with any previously employed architect that you are to be employed. (This, and the next heading 'Fees' will normally be covered as part of the Building Type brief)

Fees

Establish with client in writing or as a contract (such as those issued by the RIBA) extent of duties and amount of fees and expenses, with stages of payment

Right of access

Check with client that architect has right of access to the site

Client's liabilities:

insurance

Property, including site and any buildings on the site, should be insured after purchase.

notification

The client has liabilities under the Town and Country Planning Acts to declare his intention of development to interested parties when seeking planning consent ie owner or tenant

Cost of site

To be checked against local land prices and taken into account when assessing financial feasibility. Check also for any hidden or other charges (consult solicitor)

Existing drawings

Obtain from client or solicitor any existing drawings or documents to identify site and surroundings, and any buildings, drains or services on the site. See also para 9 which lists available types of Ordnance Survey maps

3 Legal inquiries

Discuss with client's solicitors any factors which might affect development, such as those listed in col 1

Leasehold	Establish a period of lease, responsibilities on reversion and amount of ground rent and rates
Sublease	Establish conditions as above
Restrictive covenants	These may control type of use, form of building, rent, tenure and sale
Rights of way	It is possible for these to be legally terminated. Discuss with local authority
Existing approvals	Any approvals already granted for the site
Granting of wayleave:	Public utilities (listed in col 1) and others have powers in establishing wayleave
for electric cables	Above or below ground
for gas pipes	
for water pipes	
for fuel lines	

private rights	Including drainage or other services having a right of way and right of access for inspection maintenance and repair over the site
for telephone lead lines	
for other rights of entry	Fishing, shooting or agricultural access
Rights of light and air	
Boundaries of site	Establish exact boundaries and extent of site

4 Authority inquiries

County planning:	The planning officer should be consulted about factors liable to affect development, as listed in col 1. Refer for greater detail on all planning matters to *Encyclopaedia of planning law* and to the Town and County Planning Acts [Aa5] which are administered by planning authorities
zoning of area	The county development plan should be consulted to establish how the area is zoned (residential, industrial, green, belt, etc)
redevelopment	Whether scheduled or unscheduled, or part of an estate having a particular purpose

highway planning	New roads may be scheduled or there may be road widening schemes. These may create access problems
developments	Inquiries should be made about other possible developments, such as airfield construction, pipe or cable distribution, that could affect the project
Local planning:	Consult with local authority
planning permission	Likely to be sought in *outline* at this stage. Planning permission should also be obtained for change of use when *full* planning consent is applied for. General discussion should take place on likelihood of success of project and on local feelings. It should also be established in discussion whether there are outstanding planning appeals for similar projects or if such appeals have been made in the past and their result
building line	Position, if any, should be ascertained
improvement line	Position, if any, should be ascertained and any liabilities that can be imposed under law
restrictions on access	Such as road improvements, widening of streets, one way traffic systems and roundabouts. These could affect access

planning requirements:	
density plot ratio building height restrictions parking standards	Establish definition of site area in relation to permitted density. Usually local authorities are flexible on these conditions and agreement may be reached by discussion of tentative proposals
light angles	See Ministry of Housing and Local Government Planning Bulletin 5 *Planning for daylight and sunlight*
other restrictions:	Discuss with local authority means of overcoming any of the restrictions listed in col 1
scheduled as ancient monument	
scheduled as of historic interest or conservation area	For evidence of remains below ground or features or buildings existing on site
tree preservation orders	
consecrated ground	This may involve the expense of removal and reburial
statutory allotment	Significant only if council has land registered as a statutory allotment
Department of Trade & Industry	
industrial development certificate (idc) office development permit (odp)	It should be noted whether any change of legislation has occurred either nationally or in certain areas that amends

the area permitted without certificate or permit. A check must always be made on whether these will be required and the likelihood of its being granted

Engineers' and surveyors' office:

acceptable ground loadings and soil character

These preliminary inquiries do not necessarily obviate the need for a soil investigation and report but may serve as a general guide. (See para 29–36 for full soil investigation survey)

tip site

Records should indicate whether the area has been used as a tip

depth of sewers

If sewers have been laid recently information may be available as to positions of connections. Check whether any new sewer runs are planned which could help to reduce costs of connection. Some form of soil drainage or disposal must be provided and maintained by the authority. Trade effluent is the problem of the developer and may involve heavy costs if the authority will not make arrangements. The authority usually makes a charge if it undertakes disposal of trade effluent

road adoption

If access to the site is via an unadopted road (ie not maintained by the authority), heavy costs may be involved. Negotiations may be required to settle what state the road must be in to be adopted, or what proportion of cost will be borne by the authority in bringing it up to standard. In some cases a new access may be provided by the authority, complete or on a shared basis. This will depend on how anxious the authority is for the development to take place or on the influences of other liabilities

general inquiries

From this office a great deal of information may be gained relating to local conditions on such subjects as mining influences, tunnels, water table, flooding, environmental influences and nuisances and types of contractors available. Other inquiries will become apparent from the type of work envisaged.

5 Services inquiries

Information will usually be obtained from local supply companies but sometimes also from local specialist sub-contractors. Establish availability and capacity.

Electricity
Gas

Check programme of new supplies if relying on them. Costs may be incurred in insufficient capacity for purpose of project. Check that capacity in the future will not be affected by other users

Water:

Check to what height pressure is available and whether reasonably constant

fire fighting

Check whether sufficient for sprinkler or drencher installations. Pressure tanks in this connection are costly.

	There is no obligation for the water undertaking to provide piped capacity for this
reclamation	If required for cooling or other industrial purpose the water board may require costly cooling or recovery plant to be provided for water conservation
Telephone	Availability of lines. Possible delay

6 Site environment

This section deals only with a preliminary investigation of environment and climate to establish whether there are any conditions likely to affect feasibility

Character of area: agricultural, residential, commercial, industrial	Note also whether these characteristics are growing or declining. A more careful survey may be needed later, see para 17.
Climate:	Dealt with in detail in para 18–28. Only exceptional characteristics should be noted at this stage
prevailing wind	The purpose of the site should be related to any nuisance that could preclude its use (eg excessive air pollution from upwind sources such as sewage works, tanneries and cement works)

Noise sources	Eg main roads, railways, airfields, dairies, general industry and isolated industry (eg sawmills). If noise is likely to be a serious source of nuisance it may be necessary to make a more detailed study at this stage. If so refer to para 23
Vibration	From similar sources as noise. Important if delicate instruments are to be employed. May also be underground mining or such things as metal forming machines which cannot be heard
Availability of amenities:	General description, distance from site etc
personnel	For employment in new building
transport	Note expansion or reduction of services
parking	Number of places, future multi-storey parks etc
entertainment	Scope and type
shops	Standard and type. Nearest large shopping centre
schools	Size and type. Academic standards
churches	Denominations
postal services	

housing	Local authority policy. Availability and current prices of private housing
7 Physical characteristics of site	A full site survey will not normally be necessary at this stage and a careful inspection will usually be sufficient to check that there is no obvious reason against the project. (For full site survey see para 13–16)
Size	Note also room for expansion and possibilities of purchasing adjoining land for expansion
Shape	In relation to type of proposed development
Orientation	
Physical features	Excessive slopes requiring cut and fill or affecting planning levels; attractive natural features worth preserving
Type of ground	Eg clay, sand, rock etc. (For full soil investigation see para 29–36)
Land slips	
Erosion	
Flooding	Should include observations on drainage and waterlogging. (May possibly be indicated by reeds growing)

Access to site	For delivery of materials etc while building is under construction and for access by users on completion of building
Surrounding fences or walls: condition ownership rights of support	
Adjoining buildings:	Establish whether it will prove advisable for schedules of dilapidations of adjoining properties to be prepared and photographs to be taken.
condition fire risks ownership rights of support	
rights of light and air rights for maintenance	To which site is entitled from existing buildings and to which adjoining owners are entitled. Special arrangements may be required to maintain these or to negotiate compensation
services	Check that no services feeding adjoining buildings run across the site or enjoy the support on the site

Buildings and works on site:	
number, dimensions, location	
types	
conditions basements	May affect foundation design
tanks	
roads and paths	
reuse	Consider reuse of existing buildings etc and compare costs and other advantages with those of complete or partial demolition

8 Feasibility report

A feasibility report may be required at this stage for consideration by the client. See also appropriate SfB Ba4: (91)–(98) Building Type Design Guides

2 Detailed site information

9 Maps and plans

This second part of the chapter on site investigation sets out a detailed check list suitable for most building sites. It must be read in conjunction with part 1: Preliminary site information, para 1 to 8, as information is not repeated

Ordnance Survey:

Director General, Ordnance Survey, Romsey Road, Maybush, Southampton SO9 4DH. There are also regional offices: addresses of these are listed in chapter 15 or may be obtained from the Director General

maps:

Ordnance Survey sheets are subject to copyright but a licence may be obtained to make copies. Check with office whether more up-to-date maps are in preparation. Advance information is sometimes available

6in to 1 mile

Or 1 in 10,560. These maps exist for the whole of Great Britain and Ireland

1 in 2,500
1 in 1,250

Or approximately 25in to 1 mile } Exist for most of the
Or approximately 50in to 1 mile } country

1 in 500

Or approximately 125in to 1 mile. Available for some areas only

bench marks	Ordnance Survey publishes lists of bench marks and these are more up to date than the plan references
Ordnance Survey services and publications	The Ordnance Survey office offers various services; these are set out in Chapter 15
Aerial photography:	
government	Aerial photographs taken for the Ordnance Survey by the Royal Air Force are available from the Department of the Environment
private companies	These hold libraries of photographs and offer survey services. Sections of photographs or complete prints can be obtained at reasonable cost and can prove very useful in the initial survey stage
Street maps	May be available from local bookshop or newsagent. These have limited use and should not be used as accurate scale maps
Local authority:	Engineers' or surveyors' department
drainage plans: soil, surface water	With information on levels and connections. Not always complete or correct

road levels	This information may be available in detail if the road is of reasonably recent construction
services	Plans may be available showing the combined positions of services. Otherwise this information is obtainable from the various undertakings
recent development	The local authority will have large scale maps or plans which will show more up-to-date information than is available from other sources
buildings on adjacent plots	Existing and proposed
road improvements	Any proposals affecting site
Meteorological maps	See Climate section, para 18–28
Geological maps	See Soil investigation section, para 29–36
Old plans	These may be traced by the client's solicitors or agent, or may be available from the local authority as prints. The authority may also permit tracings to be taken

10 Legal inquiries

See para 3 for preliminary legal inquiries

Solicitor:

land registry — If the property is unregistered or if it forms parts of a

	previously registered area, registration in the new name of ownership should be completed by the client's solicitors
easements	
other conditions	See para 3
addresses: adjoining owners adjoining occupiers existing tenant occupiers	
Agent:	
ground rent	Whether leasehold or sublease
rates	
tithes	
compensations	Check that there are no liabilities in this connection
access: permission keys	

11 Local authority inquiries

Names and addresses:	This is a more detailed list than the preliminary information listed in para 1

public health engineer building inspector medical officer of health mineral valuer district valuer	
town clerk	The town clerk is the principal officer of the local authority and inquiries addressed to him will always be dealt with or passed to the appropriate officer
lighting authorities	Sometimes delegated to parish councils
police	Find out person responsible for traffic control
petroleum officer fire officer	Normally a county official but name and address available from local authority
By-laws	Check no local By-laws in addition to Building Regulations. eg: Protection of timber against infestation
Submission dates: planning Building Regulations	
Application forms:	
Town and Country Planning Acts Building Regulations	Other application forms are not dealt with here as types of forms are related to types of development and necessarily come later in the design process rather than at the time of survey

Observations:	
character of district	
traditions	
skills	To work in local materials and traditional styles
history	Earlier uses of site may have bearing on proposed new works
types of local materials	
Bench mark positions:	If not reasonably near establish indestructible datum
Disposal:	
sewage	Type of connection: Establish whether local authority wants to make connection
surface water	
industrial waste	
garbage	
Planning:	
smokeless zone	This can affect type of fuel to be used and therefore method of heating
access	
sight lines	On road intersections. May necessitate set-back at corners but may apply only up to a limited height above ground level

Q

crossovers	Establish radii, sight lines and whether local authority wants to construct
heights	Possible aerodrome restrictions
Construction specification:	
for roads to be adopted	If to be adopted the local authority may be prepared to tender for this work or may insist on doing it.
for crossovers	Local authority will also undertake this work and requires information in advance to allow for twenty-eight days' notice to public undertaking
for foundations	It may be recommended for example that deep foundations are required owing to presence of clay etc and that use of sulphate resisting cement is advisable. Probable water table

12 Services

General preliminary information has been dealt with in para 5. Obtain now the details listed in col 1

Electricity: voltage phase cycle ac or dc **Gas:** size and pressure of main **Water:** hardness chemical composition size and pressure of mains copy of regulations recommended materials for mains	Drawings indicating positions of these services should be obtained from the relevant undertakings
Telephone	Overhead or underground. Duct requirements

13 Initiate survey

Before setting out to take the actual measurements and levels a clear understanding must be had of precisely what information is required, to what extent and in what detail

Type of survey: own survey	Chapters 9–14 See also *plane and geodetic surveying*

local surveyor	Local authority may supply list of suitable names
survey by specialist under contract	In this case a clear brief should be issued after discussions with the specialist. The Royal Institute of Chartered Surveyors, 12 Great George Street, London sw1, will advise on scale of fees and names of surveyors able to undertake the work
photographic aerial survey	Economical on larger site. Savings can be made if requirement is limited to bringing O.S. up to date
Additional information	This will be necessary for specialist buildings and a clear brief must be established. See appropriate sfB Ba4: (91)–(98) Building Type Design Guides
Specialist surveys:	
soil investigation and analysis	See para 29–36. An outline position of the building and proposed roads will ensure the most useful placing of bore holes
climatic survey	See para 18–28
noise survey	See para 23
market and amenity survey and report	

14 Physical survey of site

Preliminaries: time available access keys letter of introduction travel arrangements accommodation	
Grid of levels and/or contours	Decide how detailed the pattern is to be. Decide datum and key levels. Chapter 12
Photographs: by self } by contract }	A very useful photographic record can be compiled of the site as existing. The larger the better. The prints should be given reference numbers or descriptions to relate them to a key plan showing directions from which pictures were taken and the date recorded
Trial holes:	For small projects the items listed in col 1 and Chapter 17 are relevant. For larger projects requiring full soil investigation see para 31–36 and Chapters 17–19
local builders	The local authority can usually give names of local builders prepared to undertake this work but may not recommend any one builder

position and number of holes	Ideally this information can be given on a rough plan so that the holes are ready for inspection before the on-site survey begins (see para 15)
type	Whether dug or bored
size depth	
protection	Holes must be protected with substantial enclosures; clear instructions in writing should be given to the contractor and arrangements made for re-filling. Insurance must be held by contractor chosen
pumping	Necessity for pumping, probably dependent on water table
Survey equipment: list of items required	
transport of equipment	Check that suitable transport is available
hiring of equipment	Suppliers of surveying and drawing equipment will often hire items out
check condition of equipment	It should be in accurate and working order.

Documents:	
building plans deeds, leases, old maps	May help in establishing positions of old foundations and underground works, services etc
Ordnance Survey maps	Chapter 15

15 On-site survey check list

Site surveys should be extended beyond the limits of the area to be developed to include structures or topographical features which may affect or be affected by proposed new construction. Approximate ranges of influence of these items will vary according to the size and extent of the project. For large schemes, an indication of the range of influence is given in brackets, where appropriate, in col 2 below. Obviously these distances must be balanced against the size and complexity of the project

Location of site:	Will have particular relevance to climate (see para 18–28)
latitude	Particularly relevant to weather type
longtitude	Relevant to 'daylight hours' (see para 20)
Orientation of site: os grid north magnetic north	
Natural features: surface levels	(within 20m of site). For methods of taking measurements and levels of sites see Chapters 9–14

depressions	Note local hollows and mounds which may escape grid of levels
mounds	Note material
holes	Note water level
outcrops	Note whether blasting is practicable for surroundings and whether necessary
quarries and pits	(within 500m of site)
Water:	Note general condition (eg stagnant, overgrown and liability to flood)
rivers, canals and streams	(within 1 km of site). Note direction and rate of flow, type of banks, river bed etc
ponds and lakes	(within 1 km of site)
ditches and water courses	(within 1 km of site). Note direction and rate of flow, size, depth etc
marshes	Note extent
springs	Arrange for sample analysis of source if presence causes concern. It can be proved by colour dyes if a leak in pipe-

	work or container has occurred resulting in an apparent spring
Underground:	Note position and other details as listed
underground workings:	National Coal Board local offices and mineral valuer (see para 11) can advise Sometimes geological records will show information (see para 30)
type depth seam depth safety date of working	Establish risk of subsidence and policy adopted by National Coal Board for compensation etc
buried foundations: extent materials used	Old 1 in 500 Ordnance Survey maps sometimes give useful indication
wells: diameter depth condition safety	(within 500m of site). Note responsibilities under the Water Resources Act 1963 which controls permissions to take water from any natural source

tunnels:	
rail	
road	
GPO	
canal	
private	
tanks:	
septic tanks, cesspits, soakaways, water, fuel	Size, condition and present use
manholes and gullies:	Size and condition
cover level	
invert level	
Planting:	Note position (within 20m of site) and other details as listed
trees:	Check whether subject to preservation orders and whether planning authority is keen on retaining any trees
height (estimated)	
spread distance of roots	Desirable to know but often difficult to detect
girth (at waist height)	Or appropriate diameter

species	Basically whether evergreen or deciduous but exact identification should be attempted. Refer to *Trees and shrubs* or *The Observer's book of trees*
condition	For possible retention in project
bushes hedges garden beds and lawns	General note on species and condition
Buildings:	
plot on survey levels of solid floors locate steps up and down note basements and areas and record depths general condition: materials of value features to be retained	(within 300m of site except in heavily built-up areas). For full scale surveys of the total building, including interior, see check list in RIBA *Management handbook* sec 3.522, 3.523 and 3.524, or check list in AJ, 1960, July 28, p142–145. This present heading lists only those items that will affect the overall site. Assess if retaining walls will be required to provide support to adjoining features
Railways	(within 300m of site)
Roads and paths:	(within 300m of site)
positions levels construction and materials	

Services:

electricity: termination point gas: stopcock pit diameter of pipes water: diameter of pipes stopcock meter pit standpipes hydrants telephone: termination point check whether overhead line how terminated how supported	Position, depth and condition. Some may be taken overhead. Possibility of retaining existing services runs should be noted

fuel lines
district heating service

Drainage:

surface water:
- to combined system
- to separate system
- to soakaway or ditch
- interceptors
- oil interceptors

pipe sizes	
inverts and flow direction	
condition	
soil drainage:	
means of disposal	Main drainage, septic tank or cesspit. If cesspit establish whether local authority will empty it
pipe sizes	
inverts	
condition	Including effect of tree roots etc
inspection chamber types	Including rodding facilities
subsoil drainage:	
methods	
means of disposal	
construction	
Trial holes:	Site inspection of any trial holes dug in advance of survey (see para 14)
depth of top soil	
nature of subsoil	
water level	If possible note any variations

16 Physical boundary survey

Permission for access	Permission must be obtained from owners and/or tenants before measuring adjoining property

Measured Survey of adjoining property:	
windows external light wells vents and vent grilles supports trees projecting piers	Levels, extent and height of features that affect the site should be located by dimensions, complete with elevations of adjoining property if necessary to establish rights of light and light angles
other features	Such as levels inside adjoining site if applicable (eg where a change of level requires retaining wall construction)
adjacent roads	Type, condition and access
Ownership and liabilities:	Note types and condition of all boundaries
ownership of boundaries party walls party fence walls etc	Check with solicitor any relevant Acts relating to right of support
effect of new works on adjoining walls	If there is the slightest possibility of the new works causing damage or settlement a schedule of dilapidations should be prepared and agreed with a surveyor appointed by the owner of the adjoining property. Photographs dated and signed by adjoining owner
Party wall notices:	
London area	London Building Acts (Amendment) Act 1939 sec 44–59 [Aa5]

requires notices to be served for works to or adjacent to walls. Notice forms A to G, covering all conditions, are available from the RIBA

outside London

Notice in writing should be given of intentions to carry out work to or adjacent to walls at an early stage to enable a surveyor to be appointed by the owner of the adjoining property. A third surveyor, agreed to by both parties may be appointed as arbitrator

Features affecting access:

A note should be made of position of proposed access for contractors if it presents any particular problems, especially for delivery of large building components or plant

levels
trees
crossovers
road bends

Discussion with the local or county authority will establish (or may have established—see 'restrictions on access', para 4) the positions of ingress and egress if difficult features are present. Similarly if the site is on a main or trunk road

zebra crossings
lamp posts
street signs
bench marks
fire hydrants
cable ducts

Each undertaking should be consulted. Sometimes evidence of installations are not apparent

Purpose of access to developed site:	A note should be made of any special requirements affected by width and turning circles
road vehicles pedestrian railway canal contractor's plant and building components	

17 Local character

Character of area:	More detailed observations should be made at this stage of local character and evidence of changes taking place (such as upkeep, density, growth, etc)
coastal	Note effect of sea air on local buildings
agricultural	Land classification
residential commercial }	Smoke, dust, petrol, fumes
industrial	Chemical fumes, tar vapours

Character and condition of local buildings:	
building types	Ancient monuments of listed buildings etc
character and scale	
uses of materials	
failures	In structure or materials
Views and approaches	Careful note should be made of how the siting of the building is affected by views out of the site and approaching the site. Also of massing in relation to adjoining properties. If possible a point should be found where the site may be viewed as a whole so that due care can be taken with regard to general character and local features

Climate

Most meteorological data is collected for weather forecasting, agricultural or other applications not connected with building. Architects concerned with climatic conditions on building sites must select information which is appropriate to building design needs and relevant to the particular problem. Local sources of information should be consulted wherever possible. It is important, however, not to be reassured by local records of moderate conditions. For example, no local records of intense rainfall do not mean that it cannot occur. In such cases the general data indicating the probability of heavy rain such as the Bilham formula should be employed in preference to local information.

In some cases Building Regulations give minimum standards which must be reached in the building (eg insulation for walls and roofs) or Codes of Practice (eg CP3, Chap II: 1970 Thermal insulation in relation to control of environment) give design standards which in general architects should use in preference to any interpretation of local data.

In many other cases climatic data can only be applied to design problems by means of a specialised procedure which includes appropriate allowance for climatic information (eg heat loss calculations and admittance method calculations).

Much of the data which is available and relevant will be based on conditions at meteorological stations some distance from the site. Appropriate allowances for geographical, topographical and adjacent buildings must then be made in accordance with the section on interpretation of data below.

Climatic factors

18 Air

Movement:	
structural	Note importance of very short gusts
comfort (external)	Velocities below 5 m/s for comfort. Note also duration and velocity of winds from particular directions
heat loss	See IHVE guide. Note effect of exposure on U value
ventilation plant growth	
Temperature:	
heating installation	Minimum design temperature conventionally 0°C for southern England. See IHVE guide
heating costs	Mean temperature during heating season
summer overheating	Maximum design temperature. See BRS Current Paper 47/68 Summertime temperatures in buildings, and IHVE guide
freezing of buried pipes comfort (external) durability of materials	Consider maximum and minimum temperatures, diurnal variation and mean temperatures at various seasons
plant growth	

Pollution:	
corrosion and discolouration	Content of salt spray, dust or chimney effluent
comfort	Odour; typical pollen counts
view	Improved by clean air

19 Water

Vapour	See also combined factors, relative himidity. Actual quantity of water in air is known as absolute humidity. Degree to which air is saturated (which will vary with temperature) is termed relative humidity and expressed as a percentage. Relative humidity is relevant to design and is easily measured and recorded
Droplets suspended in air	These give rise to mist (visibility exceeds 1100 m), fog (visibility less than 1100 m), and cloud, which affect view, insulation, moisture penetration, plant growth
Precipitation: rain sleet snow hail	Affects gutters and downpipes, paved areas, drains and snow loading. Design conventionally based on 150 mm rain per hour for gutter and downpipes and 100 mm/h for drains. Note that criterion is the probability of heavy rain. See Bilham's formulae in British rainfall (HMSO, annual)
Surface run-off	Sharp slopes may give rise to unexpected surface flows which may be embarrassing to a designer

20 Sun

Direct radiation:	
sunlight	Desirable for human comfort. May cause glare which should be avoided in some situations, eg factories. Can affect plant growth. See Sunlight & daylight (DOE/Welsh Office, HMSO)
heat	May cause summer overheating. If problem is likely, screening may be desirable coupled with careful consideration of fenestration and orientation and thermal properties of the fabric. For calculation method see Air, Temperature: summer overheating (above)
ultra-violet	Deterioration of material and building contents. Should be totally excluded from museums and similar institutions
indirect sunlight (daylight)	Critical daylighting conditions in this country. See 'Sunlight and daylight'; BS CP 3: Chap I Lighting; Chap (I) B Sunlight; and IES Code. Glare may have to be considered in some cases

21 Radiation to cold night sky

Dew (externally)	See BS CP 3: Chap II
Condensation in walls and roofs	

22 Lightning

Severity and incidence of storms

23 Noise

Note intensity, duration, frequency and noise spectrum leading to discomfort or disabilities by occupants. See BS CP 2: Chap III Noise reduction in building

Noise levels

Establish surrounding noise levels for the most disturbing noise by measurement or use of tables. Noise measurement should be carefully planned and preferably carried out by an expert

Existing noise sources:

road traffic	Note any steep gradients which may increase noise
hovercraft	These produce about as much noise as aircraft
water traffic	eg Speedboats and diesels
aircraft	Whether low or high flying, landing and taking off, within area of sonic bangs etc.
railway traffic	Above or below ground
pedestrian traffic	Can reach high levels, especially in confined spaces
industrial noise	Note especially type of noise (eg impact) and time of noise (eg daytime only) and whether reduced level at night but still noisy

social activities	Recreation, sports stadia, public meetings, dance halls, school playgrounds, outdoor swimming pools etc. Note particularly whether sound amplification is in use
building construction road construction	Likely to be a temporary nuisance only but some large scale construction works can extend over considerable period
Future noise sources	Examine possibility of future noise nuisance, particularly road and air traffic noise and likelihood of large scale construction works over extended period
Noise nuisance to adjoining sites or properties	Check that noise being produced from proposed building will not lead to complaints about noise nuisance from neighbours. See Noise Abatement Act 1960. Industrial noise nuisance is especially likely to become a serious problem where night shifts are worked, particularly if impact noise is involved

24 Combined factors

Wind chill	Affects heat losses (air temperature), comfort externally (air movement)
Driving rain and mist	Basic factors are precipitation and air movement. Affects run-off on walls and windows, wetting of walls, penetration of wall and window joints. (See Index of exposure to driving rain, BRS Digest 127)

Driving snow

Basic factors are precipitation, and air movement and temperature. Affects penetration of roof and other joints

Snow drifting

Basic factors are precipitation and air movement: in particular local reductions of air velocity due to obstructions which allow snow to settle. Affects accumulation of snow with consequent difficulties of pedestrian and vehicular movement, and possible moisture penetration during melting. (Note also snow slipping from roof)

Relative humidity

Basic factors are absolute humidity and air temperature. Affects deterioration of materials, condensation and thermal comfort

References

25 Sources of information

Local meteorological stations can be a valuable source of information and advice *provided that the inquirer has some idea of what he wants to know*. Other possible sources of local knowledge include the local authority; architects and builders; schools and colleges which have made a study of local climate; river boards and so on. Apart from these sources, information of general application is available from the references quoted in the bibliography.

See also AJ Handbook of Building Environment, especially Section 1 Climate and topography (AJ 2.10.68 and 9.10.68)

Interpretation of data

26 Location of site

In the British Isles there is a general tendency for conditions to become colder towards the north and drier towards the east but the effects of altitude and distance from the sea will often override the general tendency. For the site under consideration, therefore, examine the likely effects of local climatic factors, topography, adjacent buildings and physical features as indicated below.

27 Topographical factors

In cases where specific records for the site are not available consider the data obtained from the nearest Meteorological Station in the light of the topographical variations between the site and the recording station. The factors listed in col 1 are relevant; the order in which they are listed is a convenient one for reviewing typical situations

Ground surfaces upwind

These will influence air masses arriving at the site, eg loss of moisture by precipitation over high ground or heat gain resulting from passage over warmer land

Altitude

Increased altitude has the following general effects:

temperature decrease;
higher wind velocities;
higher rainfall (but see 'slopes and altitude variation' below);
more snow and longer snow lay;
more fog;
reduced air pollution;
greater intensity of insolation
greater radiation losses;
greater range of diurnal temperature

Slopes and altitude variations

These influence climatic factors as indicated:

wind and rain

Slopes facing winds have greater wind velocities and these increase as the altitude of the slope increases;
slopes facing rain-bearing winds receive more rain, and the rainfall increases as the altitude of the slope increases;
slopes facing away from the direction of rain-bearing winds usually have a lower rainfall.
Note that shelter from the prevailing wind may leave the site open to storm winds if these typically come from a different direction. Sheltered sites with low wind velocities will tend to be areas where snow accumulates

sun

South facing slopes gain significantly more solar radiation. High surrounding ground may determine the distribution and duration of insulation which itself will influence snow lay and the rate of evaporation from and drying of surfaces

Areas of negative relief (areas enclosed by higher ground)

These areas are subject to low minimum temperatures and to more frequent frosts resulting from cold air flow and adverse radiation balance. Since drainage is often restricted, high humidity is often found which, combined with low minimum temperatures, may lead to higher incidence of frost and fog

Areas near sea

The distance inland subject to conditions prevailing at the sea depends on local topography. Generally, the flatter the area the greater the penetration but on hilly or mountainous coasts the effects may be funnelled inland. The following should be considered up to 8 km and in suitable conditions up to 16 km from the sea:
sea mist;
land and sea breezes (onshore by day and offshore at night);
wind blown sand;
increased salt content in air;
reduced range of diurnal temperature

Nature of soil

Surface and air temperature ranges will be greater over soils of low *thermal diffusivity*. The greater the moisture content of the soil the higher will be its thermal diffusivity and the lower the temperature ranges experienced

Urban areas

Urban areas have the following special characteristics:
increased cloudiness;
increased precipitation;

greater atmospheric pollution in the air down wind of the area;
higher minimum and mean temperatures;
reduced humidity resulting from efficient drainage preventing return of moisture to air by evaporation. (This reduced evaporation contributes to the higher mean and minimum temperatures); significantly more turbulent air movement

28 Buildings and adjacent physical features

Buildings, trees, hedges and walls are small scale topographic features that will, locally, exert influences similar to that of major topographic features. The following effects may be discovered:
increased wind velocity or shelter due to form of physical features;
shading from the sun which may affect snow lay and surface moisture;
buildings, walls and so on may have the effect of areas of negative relief so that lower temperatures and higher humidity may result.
In addition effluent from other nearby buildings and disposal of effluent from buildings should be considered

Heat storage

Many buildings, walls and civil engineering structures have a greater thermal capacity than the natural materials that they replace, and building surfaces are often better drained. More solar heat will therefore be stored to be released during the night. The extent of the effect will be governed by the capacity and size of the building elements and the duration and timing of their insolation. In suitable circumstances the effect may be significant

Air movement

Tall buildings themselves affect the immediate environment.

Soil mechanics site exploration

This work, and the interpretation of results and foundation design, should normally be supervised by a specialist engineer. The Institution of Civil Engineers, 1 Birdcage Walk, London sw1; Associated Consulting Engineers (ACE) Ltd., 19 Collingham Road, London sw5 and the British Geo-technical Society (c/o the ICE at the same address) will recommend the names of suitable engineers.

Soil Investigation: General

The design and construction of foundations and other below ground works may, according to circumstances, be quite straight forward or present such problems that an extensive soil investigation is essential. But except in the very simplest cases a reasonably exact knowledge of the soil conditions makes it possible to reduce the cost of a design without loss of safety, and gives the designer greater certainty of how the completed structure will 'behave'.

Large and complicated investigations for a small project are rarely justifiable or indeed economically possible, but use of the general body of foundation theory available, plus a small scale site investigation, can increase the efficiency of the foundations.

A site investigation adding in most cases only one per cent or less to the total project price will normally suffice to give adequate information. The total cost of the investigation (excluding any specialists' work) can range from £8 to £30 per metre of bore hole depth, but is likely to be in the region of £15 per metre for the average project. Such a cost is well within the normal amount allowed for contingencies, and in all but a very few cases the cost will be covered by savings in construction, for the use of test figures in the calculation of allowable bearing pressures will give margins of saving over rule-of-thumb methods. These can be misleading and are frequently too conservative. If the site investigation information is available to the contractor, uncertainties (and possibly later claims—most often encountered for work in the foundation stage) can be reduced at the tender stage. Obviously if no information is available on the materials in which the foundation is to be executed, then every circumstance in the construction is unforseen. Information on ground water conditions alone will be well worth the small cost involved in exploring the site strata.

The selection of the investigation work to be carried out and its actual execution must be done by experienced persons. Unnecessarily elaborate investigations are nearly as bad as none at all. A clear idea of the proposed project is essential for a soil investigator; and an awareness of the information needed for economical foundation design is required.

29 Check necessity for soil investigation

This applies to both new projects and to extensions (eg heightening) of existing structures.
A site investigation may not be required if sufficient is known about the site from the local authority sources, previous or adjacent site investigations, or from local knowledge, or if the building is of a simple type producing very light ground loadings eg less than 25 KN/m^2
Alternatively in such cases the work might be confined to a very simple confirmatory investigation eg a few trail pits or short hand-auger bores
Some form of investigation, either small or large, is normally required if any of the following conditions apply:

Proposed site loading

If new building is of other than very light construction

Type of site

If site is of peat, marsh, fill, subject to flooding or recently reclaimed

Ajoining sites or properties

If site is marine, estuarine or adjacent to river or to existing buildings

30 Examine existing geological records

Refer to geological records which show upper drift deposits and/or underlying rocks. These are obtainable or viewable at the Geological Museum, Exhibition Road, South Kensington, London sw7 and at regional offices of the Geological Survey, and local libraries.

Also published are regional geological memoirs by the Geological Survey. Local geological society records, where they exist, can often be useful. However, both maps and memoirs are small scale and can give only a general indication of the type of soil to be expected.
No information concerning physical properties is given.
See:
Good Soil

31 Decide outline of type of soil investigation required

Trial pits

Usually sufficient for small areas and loads. (Typically two-storey houses). See Chapter 19

Trial borings

For larger areas and loads, sometimes in conjunction with trial pits. Full requirements should be checked with the soil mechanics engineer. The spacing of bores (or pits) over the site will depend on the particular circumstances, but will normally be between 15m and 50m centres. Not all need necessarily be taken to the full depth indicated by para 34. See Chapter 17

Testing

This should be done with small or large scale investigations. Tests may be made in the laboratory on samples taken from the site, or tests may be carried out in situ, where applicable. Often both laboratory and in-situ tests will be appropriate. See Chapter 18

32 Prepare contracts for soil investigation

Advice on engineers or firms who will supervise or undertake site investigations can be obtained from:
The Institution of Civil Engineers, 1 Birdcage Walk, London sw1
The Association of Consulting Engineers, 2 Victoria Street, London sw1
Federation of Civil Engineering Contractors, Romney House, Tufton Street, London sw1

Decide how the soil investigation is to be undertaken:

by small number of shallow trial pits

Employ local resources with facilities for technical supervision and visiting

by larger investigation

Employ a specialist firm or firms undertaking borings for the trial borings on site. The laboratory testing work may be done by the same firm or by a specialist engineer or laboratory. The in-situ testing work can sometimes be done by the drilling contractor, but large scale tests may require a firm with more resources. It is preferable that the organisation which does the testing does the reporting and if possible supervises the drilling and any other site work

Form of contract, specification and bill of quantities

Except in the case of very small investigations, should be drawn up by a specialist engineer, and the contract to go out to tender either as an advertised tender or on a selected list. See Chapter 19

Contract information and procedures are dealt with in Chapter 19

The main considerations are listed below under site work and dealt with fully in this chapter

Further information is also available in:

Soil mechanics in engineering practice p.255–285

Basic soils engineering chap 15 and appendix VII para 2

Site work

The form of contract should cover all requirements for:

Size and number of bore holes

In-situ tests

Sampling

Records of all site work

Laboratory work

This may be in the contract or may be dealt with separately in another laboratory. Laboratory work should proceed concurrently with site work to enable further samples from bores or pits to be obtained if laboratory testing shows the need for them. Check need for further testing in conjunction with specialist engineer and/or contractor

See: *Basic soils engineering* P.454 et seq
Laboratory testing in soil engineering

33 Contents of soil investigation report

The report may be prepared under the site investigation contract or may be prepared by the engineer who carried out the laboratory work. In any case the report should be prepared by a person familiar with the testing and site work and preferably also familiar with the proposed structure on the site

Details of work carried out under contract and in the laboratory

The report should include details of the site exploration work, in-situ testing and laboratory testing, or either fully or in part by a consulting engineer. Chapter 19 shows the recommended form and contents of the report

Recommendations

This part of the report, based on the known requirements, should be submitted by either the engineer or specialist contractor, or both. But this part of the report also implies information being given by the architect or structural engineer on the area and height of the proposed buildings, proposed form of construction; loadings transmitted to the ground and an estimate of allowable settlements

The need for further investigation and tests may arise on a site where the comparatively small-scale tests completed in bore holes and trial pits, combined with laboratory tests, are insufficient to give design information owing to site conditions requiring further investigation or testing to overcome extrapolating errors; eg the presence of many small random lenses of soft material embedded in firm strata may require full scale tests to provide bearing capacities and settlement data. This may not be known until the final site investigation report is prepared in conjunction with the structural designer. In such a situation a further site investigation will be needed. This will however need to investigate only the areas of ignorance shown in the first report. See Chapter 19

Copies

Copies of at least the factual section of the report should be available in due course for all tenderers for the foundation and building works
(Information used in the design of permanent works is the same as that required for the design of temporary works)

34 Use of soil data as project information

The use of the data supplied in Chapters 17–19 will provide information for the design of the elements listed below. Listed after each element are the main items of soil data necessary for the design of that particular element

Foundations:

For design of type, size and depth of foundation required

Soil data required:

soil type
soil density
soil moisture content
soil strength
soil consolidation properties
water table position

All to a depth below founding level of at least 1.5 times the likely breadth of the largest foundation where approximately square in plan, but up to 3.0 times in the case of long strip footings. If isolated foundations are more closely spaced than about three times their breadth centre-to-centre they should be considered, like a raft, as a single foundation for this purpose. When piling is envisaged, the proposed pile length should be added to this depth (calculated from the breadth of the pile groups) to give the measurement from the top of the piles

If a known hard stratum of sufficient thickness occurs, the depth may be reduced, but it may have to be increased if there are appreciably softer underlying strata.

See:

The AJ Handbook of Building Structure Section 4 (1972)

Retaining walls:

For design of type and size of required retaining walls

soil data required:

as for foundations

To a depth below ground of twice the height of the wall or greater if the wall is surcharged. See also the AJ Handbook of Building Structure Section 4 (1972)

Earthworks:

For design of embankments, slopes, cuts and other permanent excavations. Note: the long term stability of slopes, especially in clay materials, is very complicated and needs the most thorough investigation by a qualified engineer

soil data required:	
soil type	
soil density	
soil moisture content	
soil strength	Both long and short term
soil consolidation properties	
soil compaction characteristics	
groundwater conditions	
Roads:	For design of permanent or temporary roads and runways
soil data required:	
soil type	
soil density	To a depth of about 3m
soil moisture content	
soil strength	Strength under live loading beneath a pavement, normally obtained empirically eg, measurement of California Bearing Ratio
soil compaction characteristics	For long term moisture content stability
soil suction	See also the AJ Handbook of Building Structure (1972)
Temporary works: soil data required: may be any combination of those listed above	Although the design of these is not normally the responsibility of the architect he should be able to satisfy himself that such designs are sound. They may however be subject to approval by the engineer. Design of temporary works for construction purposes may include temporary retaining

walls; strutting; temporary cuts and embankments; quantities of ground water to be disposed of during construction; foundations of cranes, derricks and tower cranes; temporary bridges or ramps

Waterproofing:

For design of precautions against water pressure, uplift, chemical attack and flooding

soil data required:
- ground water conditions
- chemical analysis of water and soil
- permeability of soils

Drainage:

Design of permanent and temporary drainage
Temporary drainage (normally during construction) may include well points, deep and shallow wells, cut-off walls. The point of disposal should be chosen with care, so that the stability of the permanent works and adjacent buildings and services is not impaired

See also:
The AJ Handbook of Building Service and Circulation: Section 4 Drainage and Sewage Disposal (1969)

This covers possible movements other than movements caused by the loading of foundations or the design of earthworks such as barks or cuts.

35 Special earth movements

Examples:
Heave caused by excavation
Swelling caused by frost entering ground
Shrinkage caused by heating of ground
Shrinkage caused by vegetation
Swelling caused by removal of vegetation
Movement caused by drainage
Movement caused by mining works
Movement resulting from geological formation (eg landslides, shifts or flows)

The type of soil data needed for each design problem will probably be unique for each case. A specialist should be consulted if necessary.
If coal measures occur under the site preliminary advice about past, present or future workings can be obtained from the Divisional Offices of the National Coal Board, but independent mining engineering and/or legal advice may also be required.

36 Influences of climate on soil investigation

A different approach to soil investigation may be necessary in climates very different from those in Great Britain

Part 4
Soil investigation

Chapter 17
Soil investigation: Methods of boring and sampling

This chapter describes drilling methods, sampling and the tools used.

General

The purpose of boring is to reveal cross-sections of the ground soil or rock strata to whatever depth is necessary. The method used must reveal as large a cross-section of the soils as possible with minimum disturbance.

For shallow depths where ground water is not present, trial pits achieve these ends more successfully than boring. Where pits cannot be dug because of depth, lack of time, presence of ground water or hard strata, then bore holes must be drilled. The tools used in making a circular vertical hole from ground surface downwards may be operated by hand or by machine. The types of tools vary according to the nature and hardness of the soil.

Table I Drilling methods for hard and soft strata at increasing depths

Depth	Type of soil strata Hard	 Soft
Very shallow: up to 3m	Trial pits or hand drilling	Trial pits or hand drilling
Shallow: up to 6m	Machine drilling	Hand or machine drilling
Deep: more than 6m	Machine drilling	Machine drilling

Even for shallow work, drilling is normally done by machine as it is quicker and only about 50 per cent more expensive, say about £4·50 per m depth drilled as against £3 per m for hand drilling. Hand drilling methods normally use auger tools on rods, the augers being simple screw type or post hole type. It is possible to use more sophisticated tools, but the extra work is unjustifiable when drilling can be done by machine. Hand augering can often be a cheap and useful check on ground conditions below the bottom of a trial pit.

Machine methods of drilling bore holes can be divided into those soft strata and those for hard strata or rock.

Methods for soft strata

The rig

The method used for soil investigation for structures is known as shell and auger boring. With machine methods, heavy tools are attached to ropes or rods suspended from steel tripods. The rope, rods and tools are moved vertically in the hole by a power operated winch **1**.

The tools

There are three main types of tool: for penetrating clay; for penetrating sands and gravels; and for penetrating harder strata such as shales and sandstones. Subsidiary to these are the 'fishing' tools for retrieving tools lost or damaged in bore holes **2**. When a tool is filled with soil it must be lifted from the bore hole and emptied before reuse. Some of the penetration is carried out by rotating the tools; this is a hand operation and is, therefore, slower.

Clays are more slowly penetrated than sands and gravels, especially when hard. If much laminated and very hard, augers may slip and the addition of water to the bore hole may be necessary if the cutter is used. The water must be removed before sampling, but does not normally interfere with the obtaining of good samples.

Sands or gravels are penetrated very quickly by the

1 *A modern drilling rig*

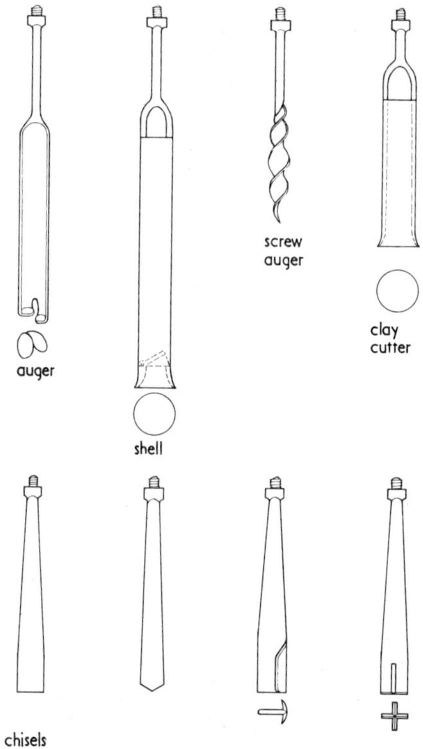

2 *Tools used for drilling 'soft' strata*

Table II Types of tools used to penetrate different strata

Strata type	Tools used	How attached to rig	Type of movement used to penetrate and fill tool
Clay	Cutter	Rope or steel wire rope	Up and down
Clay	Auger	Rods	Rotation
Sands, gravels	Shell	Rope or steel wire rope	Up and down
Sands, gravels (when very compact)	Screw	Rods	Rotation
Harder strata	Chisel shell	Rope or steel wire rope	Up and down
Harder strata	California bit followed by shell	Rope or steel wire rope	Up and down

'shell' which is a simple lift pump with a 'clack-valve' at the bottom. This cannot work unless it is full of water so that water has to be added to the bore hole if dry sands or gravels are being penetrated. Care must be taken to avoid washing fines which produces false samples.

Harder strata have to be penetrated by two proces-

ses; breaking out and then 'shelling'. Therefore the boring rate is much slower and more costly.
Drilling tools are made in various sizes, described by the nominal diameter of the bore hole they drill. For soil investigation work the size can occasionally be as small as 50mm or as large as 900mm, but the most usual size is about 150mm in diameter which is a suitable size for sampling and economical for casing.

Support of bore holes

Bore holes are frequently drilled through soils which collapse immediately the tool has penetrated them and removed its full load. Such soils are sands, gravels, silts and soft clays.
As a support bore hole casing is used. This is high tensile steel tube in lengths of about 2 to 3m, made in diameters which allow the tools to fit closely inside and work up and down or rotate. The casing also prevents entry of ground water into the hole. A certain amount of water may be necessary for drilling, but too much may affect the sampling and will also tend to cause the sides of the hole to collapse.
The casing is driven downwards, just ahead of the tools, by a hammer operated by the same winch as the tools, lengths being added as the hole deepens and joined by flush threaded male and female screwed joints, which have no protuberance greater than the outside diameter of the casing. This allows relatively easy removal of the casing, either by the winch or by jacking.
If a bore hole penetrates firm clay, no casing may be necessary. If, however, a lower stratum is reached which requires casing, then the whole depth has to be cased, even though the top part of the casing is not necessary. When water enters the bore hole faster than it can be removed—or baled—by the tools, then the hole must be cased. Casing is expensive and heavy, and is easily damaged. The cost rises rapidly with the diameter, therefore bore holes are kept as small as possible consistent with obtaining good samples and with the depth required. The deeper the bore hole the larger the diameter required at the higher levels. This is to compensate for non-verticality (which may quite well be 200mm in 6m), deflections in the hole, and to prevent tools sticking in the bore. The size of the bore hole decreases to the lowest level **3**.
It is very unusual for site investigation methods of drilling to be used below 80 to 90m. If they were the work time taken to lift and lower the tools would be excessive and problems of verticality would be difficult to overcome.

Table III General guide to bore hole sizes for different depths

Total depth of bore hole	**Starting diameter of bore hole**	**Final reduced diameter of bore hole**
Up to 25m	150mm	150mm
25m–45m	200mm	150mm
45m–60m	250mm	150mm
60m–80m	300mm	150mm

Note—In practice the sizes are arbitrary

Samples obtainable during 'soft' boring

The three types of samples are: disturbed samples, ground water samples and undisturbed samples. Similar types of sample are obtainable with hand methods from trial pits and hand drilled holes.

Disturbed samples

Disturbed samples are obtained as drilling proceeds and are samples of the soil taken from the drilling tools when they are lifted out to be emptied. They can be small or large. Small samples normally weigh about 0·5kg and are kept in air-tight containers (frequently Kilner jars) labelled by the driller with the depth from which they came, the soil type, the date and so on. For clay soils, such samples will give a general indication of soil type and moisture content. For sands and gravels they

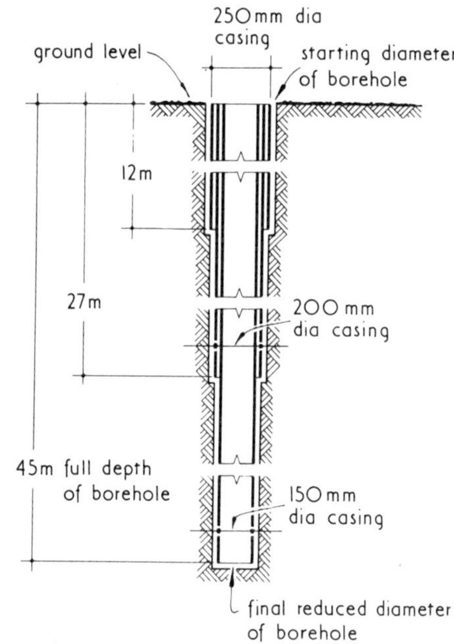

3 *Casing reduces in diameter at depth*

may well be misleading on both counts because small samples are not representative. Large samples are normally necessary only for sands and gravels, to obtain a representative quantity, and may be up to 25kg in weight. They will necessarily be taken over a range of depth in the bore hole to make up to the quantity. In view of the loss of fines etc associated with the taking of such samples their use is limited.

Disturbed samples of harder strata are of use only for soil or geological classification for type.

Ground water samples

These are taken for chemical analysis of the water. If it is impossible in drilling to separate water tables where more than one occurs in a hole, the samples should be taken after the hole is completed.

If water has been added to help in boring, then as much as possible should be baled from the hole after completion, and the hole allowed to stand for at least twelve hours before the sample is taken.

The sample should be obtained by lowering a clear container, weighted to sink below the water level. A quantity of about one litre is usually sufficient. In a cased bore hole it may be necessary to lift the casing to obtain a water sample if the water has been effectively sealed off. Long term observations of ground water movements are often desirable and may be made if a perforated tube is left in the bore hole.

Undisturbed samples

These are samples designed to extract portions of the soil which as nearly as possible maintain the

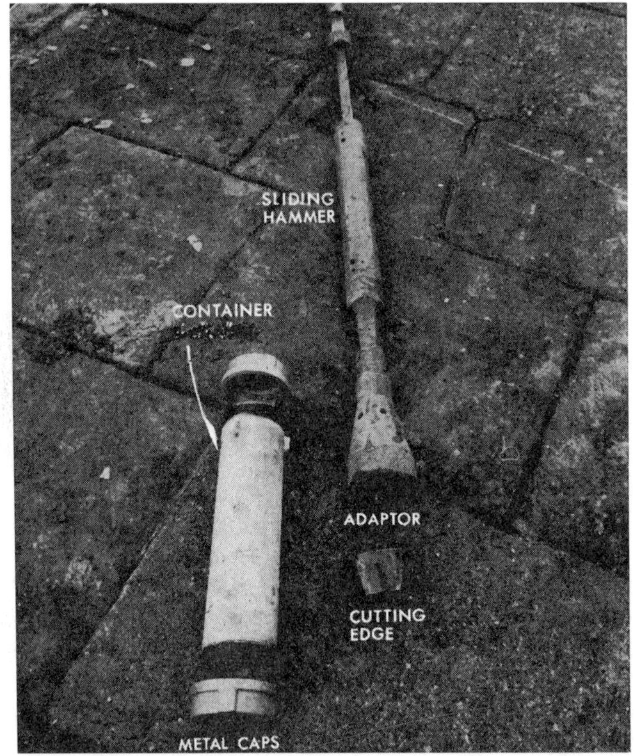

4 *Undisturbed sampling apparatus*

properties of the soil as it existed in-situ.
It is possible to obtain such samples in clays and silts and—more rarely and with difficulty—in sands. It is not possible with gravels. In trial pits, undisturbed samples may be obtained by hand cutting. Such samples require very long and laborious work, and most samples are obtained by methods similar to those used in bore holes.
In a bore hole, the undisturbed sample is obtained in a cylindrical container which is driven into the soil at the point to be sampled. The container has a removable cutting edge, and is normally attached to a sliding hammer which is operated from the ground surface by rods. Usually the hammer is hand operated. The container is designed to be as thin as possible to minimise soil disturbance, and is either steel or aluminium alloy.
If soils to be sampled are very sensitive to disturbance (ie if disturbance will greatly alter their properties), the container may be jacked into the soil. Several such samplers, called piston samplers, are available. Their use is normally confined to very soft silts and clays.
Some soils, such as fine sands, tend to fall out of the container when it is removed. For such soils a bishop sampler may be used; this has an air-tight seal at the top so that air pressure holds the sample in the tube.
The great majority of undisturbed samples is taken with the hammer-driven type of tool. The normal length is about 450mm and diameter 100mm. After driving, the container is rotated to break the bottom soil plane and the container is lifted from the hole. The cutting edge is removed, the container sealed top and bottom with wax and metal caps, and labelled with the depth of sample. The sample is then sent to a laboratory for testing. Such a sample gives a good idea of the density and moisture content and strength of a soil.
Good sampling is difficult. The main errors are:
overdriving the container,
driving into disturbed soil,
not clearing the borehole before driving,
using containers which are corroded on the inside and therefore disturb the soil by friction,
disturbing the soil when extracting,
taking a sample in a bore hole where the casing has been driven too far ahead of the tools.
Certain soils require very many blows of the hammer to drive the container, eg marls may require 160–180 blows. Obviously samples of such soils are more disturbed than those which require fewer blows. Therefore the number of blows should be recorded as part of the sampling data.

Methods for hard strata

The tools

When soft boring becomes too slow or impossible, the rotary drilling method must be used (the term 'rotary drilling' is reserved for rotation by machine).

There are only two main types of drilling tool: full-hole tools or bits, and coring tools or bits. Fishing tools are used to retrieve damaged gear.

Full-hole bits are rotary bits which bore out the full diameter of their nominal size. They are usually tungsten or hardened steel with various cutting edges. Their use is normally confined to penetrating strata from which samples are not required.

Coring bits are cylinders with cutting edges which cut a core of the strata penetrated. The core slides up inside the middle of the rotating bit as it penetrates the rock strata. Coring bits usually have a cutting edge formed with industrial diamonds, but there are bits which use shot as an abrasive. Shot drilling is slower and is normally used only in large sizes because of cheapness compared with diamonds.

The bits rotate at quite high speeds, up to 1,500 rev/min. They are attached to rods which must be lengthened as the hole deepens, pressure being exerted on the bit at the bottom of the hole to make it bite. The bits are cooled as they rotate by water or air delivered down the hollow connecting rods. In the process of drilling, casing must often be placed to support the upper strata.

The drill is designed for these several functions and is both much more complicated and much more expensive than a soft boring rig. This, and the high cost of diamonds, accounts for the higher cost of rotary drilling over soft boring (anything from £10 to £45 per metre, depending upon depth and strata, against £4 to £15 for soft boring).

Drilling

The techniques of rotary drilling are complicated and require great skill. The type and size of bit, speed of rotation, pressure on the bit and water pressure are all controllable variables to be suited to the strata penetrated. Generally speaking, the most difficult formations to penetrate are fractured rocks which break in the bore hole and damage and jam the bits. Softer strata and massive uniform rocks are easier to drill, but the harder the rock the slower the rate of drilling.

In soil investigation work on a site where soft strata overlie hard strata, it is common to use both soft and rotary methods. The hole made by soft methods is cased, and rotary drilling begins where the soft drill stopped. Rotary drilling is used to penetrate the soft strata only when information is not required on the overlying strata, since full hole bits are used with which useful samples cannot be obtained.

All rotary boring relies on a good 'flush' of the bit, either by water or air, both to cool the bit and remove the soil particles cut by it. Where water or air is lost through fissures, the hole has to be cased. An alternative is to use 'mud-flush' which—at its simplest—is a suspension of bentonite in water. This provides support, blocks fissures around the borehole and carries away larger soil particles than water flush. For very deep holes this is the only method of keeping the hole open. Specification of drilling muds is a specialised science.

In all rotary work the size of the hole drilled is only slightly greater than the bit used, which keeps water losses down and water pressures up. As in soft boring, a deep hole is started with larger diameters and is reduced with depth.

Sampling in rotary drilling

The only samples obtained in rotary drilling are the cores formed in the hollow centre of the coring bit. To collect these, a tool known as a core barrel is fixed to the top of the coring bit, with the drilling

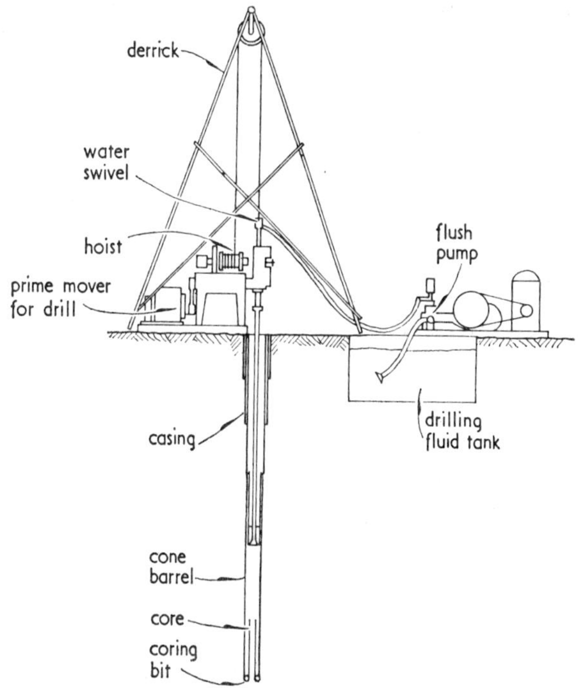

5 *A full-equipped rotary rig*

rods attached to the top of the barrel. The barrel, which is usually about 2m long, has springs inside to hold the core as it rises. When the bit has penetrated a depth equivalent to the length of core barrel, the whole assembly is lifted and the core extracted for examination.

The type of core barrel is modified to suit differing formations, the two main types being single-tube barrels and double-tube barrels. Single tubes are used for homogeneous formations and double tubes for fractured and soft formations.

The core obtained from such drilling gives a continuous record of the rock strata penetrated; and the rate of wash water indicates the degree of fracture of the rock. Great care must be taken to make sure that the depths of coring are recorded and compared to the length of core brought to the surface. The ratio $\frac{\text{length of core extracted}}{\text{length of core drilled}}$ is known as the recovery rate. The softer and/or more fractured the rock stratum, the lower the recovery rate. In good homogeneous rocks, the recovery rate should be 100 per cent.

Table IV General guide to the rate of drilling soft and hard strata

Strata	**Method of boring**	**Depth of working (m)**	**Range of daily total depth drilled including sampling per 8-10hr day (m)**
Clay	Uncased—'soft' boring	0–30	18–30
Clay	Uncased—'soft' boring	30–75	12–25
Clay	Cased—'soft' boring	0–30	12–18
Clay	Cased—'soft' boring	30–75	6–12
Sands and gravels	Cased—'soft' boring	0–30	9–15
Sands and gravels	Cased—'soft' boring	30–75	6–12
Soft rocks (limestone etc)	Rotary	0–90	8–10
Medium hard rocks (sandstone, dolomite etc)	Rotary	0–90	5–8
Hard rocks (granite etc)	Rotary	0–90	3–6

Chapter 18

Soil investigation: In-situ and laboratory tests

This chapter classifies the common tests, describes the procedure and indicates the uses of the results

Class of test	Purpose	Names of tests
1 To identify types of soils	A necessary preliminary to further tests	Visual inspection Wet or dry sieving Sedimentation test
2 To establish physical, mechanical and chemical properties of soils	To find properties which influence the strength, behaviour under load and chemical reaction when in contact with building materials	Tests to determine: natural moisture content bulk and dry density maximum and minimum densities chemical and index tests
3 To establish strength of soils	General design of foundations and soil retaining structures	Unconfined compression test Triaxial compression test vane test Menard pressiometer Standard penetration test Dynamic or cone penetration test
4 To establish behaviour of soils under a specific type of loading or stress, or with time	General design of foundations	Consolidation test triaxial compression test dynamic or cone penetration test plate bearing test full scale pile test
	General design for roadworks and earthworks	California bearing ratio test compaction test
	To discover permeability of ground for drainage	Constant head permeameter test Falling head permeameter test Pumping test
5 Tests for stabilised soils	To control quality of soil removed for use in new position	Tests to check: moisture content dry density air voids California bearing ratio pore pressure tests

CLASSIFICATION OF TESTS

Description

There are five classes of test:

1 tests to identify types of soils;

2 tests to establish the physical, mechanical and chemical properties of soils;

3 tests to establish the strength of soils;

4 tests to establish the behaviour of soils under a specific type of loading or stress, or in the course of time;

5 tests for stabilised soils (ie soils for constructional use).

Classes 1 and 2 are usually made in the laboratory and classes 3, 4 and 5 are made partly in the laboratory, partly in situ. Generally in-situ tests are more expensive; their greater accuracy is not always necessary. In-situ tests are not to be confused with 'tests on site', which are laboratory type tests carried out on the site.

The results of classes 1 and 2 are required in all investigations to decide whether tests in classes 3 and 4 are necessary.

Laboratory testing

The tests are made on disturbed or undisturbed samples of soils which have been obtained by methods described in Chapter 17. The properties of the soil may be altered by the testing.

In-situ testing

The tests are made on the soil in its original position. The soil immediately next to the soil being tested may be changed or removed but the test result applies to soil which has not been removed from its original position, and, as far as possible, has not had any of its properties altered before or during the test.

Note on rock

From the foundation engineering point of view, rock (which may be defined as a strongly cemented natural deposit) is regarded as a special kind of soil. Some softer forms of rock (eg weathered shale and chalk) approximate to normal soils, and a number of the tests described later may be applicable. Massive hard rocks however, are the subject of the science of rock mechanics, and although compression tests on small samples may be of use for classification purposes, their strength in relation to the support of heavy or concentrated foundation loads depends mainly on the extent and disposition of features such as beeding joints, fissures, faults and weathering. If loads are such that the effective strength of 'bed-rock' is an important consideration, it will normally be necessary to obtain specialist geological and engineering advice.

DESCRIPTION OF TESTS

Class 1 Tests to identify types of soils

The main types of soils (ie natural deposits excluding solid rock), listed in order of size of particles, are:

Boulders	large
Cobbles	↓
Gravel	
Sand	
Silt	
Clay	
Organic	small

There are innumerable combinations of these main types, and there have been several attempts to provide definitions of the various combinations for easy reference (see table 1). In practice, verbal description is best (see BS CP 2001 and BS 1377).

Visual inspection

With practice, this is a very accurate technique, especially in soils with particles larger than those in silt.

Table I Types of soil combinations

	Type	Nature
Coarse-grained soils	Gravel and gravelly	Well graded, fairly clean Well graded, clay binder Poorly graded, fairly clean Excess of fines
	Sands and sandy	Well graded, fairly clean Well graded, clay binder Poorly graded, fairly clean Excess of fines
Fine-grained soils	Silts, very fine sands, rock flour	Low compressibility. Silts, rock flour, clayey fine sands. High compressibility. Micaceous or diatomaceous fine sandy and silty soils
	Clays (inorganic)	Low compressibility. Medium plasticity. Sandy and silty clays High compressibility. High plasticity. Flat and heavy clays
	Clays (organic)	Low compressibility. Organic silts High compressibility. Organic clays : medium high plasticity
Fibrous organic soils	Peat and other highly organic swamp soils	Very high compressibility

Soil descriptions should include colour, odour where applicable, structure (eg homogeneous or stratified) and an indication of consistency or density. Simple field tests for soil identification are given in Table II taken from the Building Research Station Digest 64 (Second Series). An approximate check on clay strength may be made with a 'pocket penetrometer', which is a small spring-loaded plunger that records the force just necessary to push it a fixed distance into the clay.

'Made ground' or filling may include any of the natural soils and also man-made material, but will usually be obvious by reason of its variability or relatively poor compaction.

Wet or dry sieving

In this test the sample is passed through sieves decreasing in size (see BS 1377). Sieve apertures given in BS 410 vary from 125mm down to 0·075mm. The BS seive sizes are not metricated, but the previous imperial-based sizes are still, in use and the references on graph 1 are to these, pending the publication of a new grading chart in the metric version of CP 200L which is expected to be issued during 1972.

The percentage of the total sample held on each sieve is plotted on a graph **1** with the sieve aperture at the top and the soil type at the bottom. It will be seen that soil which passes the 200 (0·075mm) sieve is silt and clay. These cannot be sieved and if the proportions of these are required, the sedimentation test must be made.

Sedimentation test

Soil too fine for the 200 (0.075mm) sieve is dispersed by mechanical and chemical means into a solu- tion'. It is then allowed to settle: the faster the sediment is formed, the larger the soil particles. By means of sampling the solution at various time intervals and weighing the soil in the sample, the relative sizes of the particles can be calculated or the decreasing specific gravity of the solution can be measured. From this test, the graph to the left of the 200 (0·075mm) sieve line can be completed.

Description of soil by type

From these tests, a soil can be described by type. Mixtures are described according to the types in the mixture eg 'gravel with some sand' or 'sandy gravel'. Usually the lesser part of a mixture is used as an adjective, eg 'sandy' or 'clayey', when 20 per cent to 40 per cent of that type is present. Below 20 per cent, the words 'with a trace' normally mean up to 10 per cent and 'a little', 10 per cent to 20 per cent. Equal

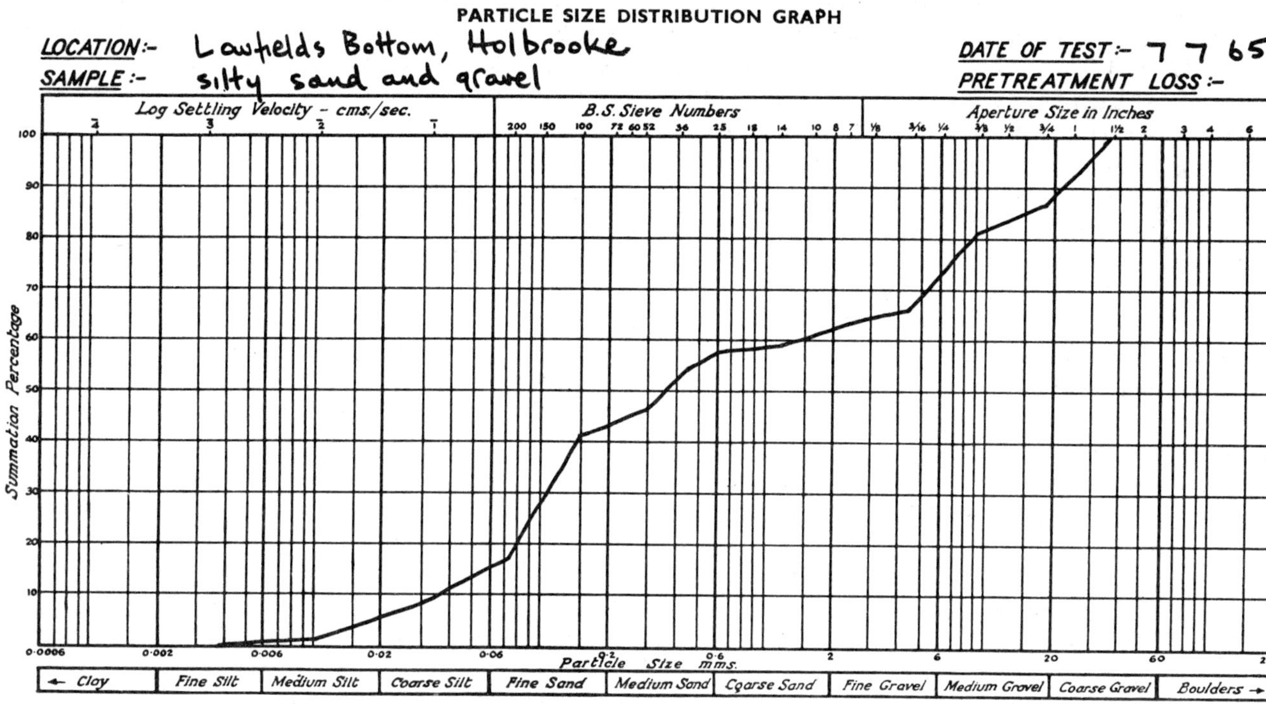

1 *Particle size distribution graph*

mixtures are denoted by using two nouns, eg 'clay and sand'. The proportions of coarse, medium and fine (for silts, sands and gravels) are denoted by a similar procedure. Where a soil covers a large range of types, presentation of a graph **1** is more useful.
Peat and organic soils can be tested only by visual or olfactory examination. Their presence is usually obvious. Properties of such soil are not normally required as it is nearly always unsuitable for use as a foundation and has to be removed, or by-passed by piles.
Accurate classification of soils is extremely useful, as it gives an idea of the foundations problems to be solved by further investigation.

Table II Soil identification

Soil type	**Field identification**	**Field assessment of structure and strength**
Gravels	Between $\frac{1}{16}$ and 3in diam. (retained on No. 7 (2·36mm) BS sieve)	Loose—easily removed by shovel
	Some dry strength indicates presence of clay	2in stakes can be driven well in
Sands	Pass No. 7 (2·36mm) and retained on No. 200 BS sieve	Compact—requires pick for excavation. Stakes will not penetrate more than a few inches
	Clean sands break down completely when dry. Individual particles visible to the naked eye and gritty to fingers	
Silts	Pass No. 200 (0·075mm) BS sieve. Particles not normally distinguishable with naked eye	Soft—easily moulded with the fingers
	Slightly gritty ; moist lumps can be moulded with the fingers but not rolled into threads	Firm—can be moulded with strong finger pressure
	Shaking a small moist lump in the hand brings water to the surface	
	Silts dry rapidly ; fairly easily powdered	
Clays	Smooth, plastic to the touch. Sticky when moist. Hold together (when dry). Wet lumps immersed in water soften without disintegrating	V. soft—exudes between fingers when squeezed
		Soft—easily moulded with the fingers
	Soft clays either uniform or show horizontal laminations	Firm—can be moulded with strong finger pressure
	Harder clays frequently fissured, the fissures opening slightly when the overburden is removed or a vertical surface is revealed by a trial pit	Stiff—cannot be moulded with fingers
		Hard—brittle or tough
Peat	Fibrous, black or brown	Soft—v. compressible and spongy
	Often smelly	Firm—compact

Soil type	Field identification	Field assessment of structure and strength
	Very compressible and water retentive	
Chalk	White—readily identified	Plastic—shattered, damp and slightly compressible or crumbly Solid—needing a pick for removal
Fill	Miscellaneous material e.g. rubble, mineral, waste, decaying wood	

Class 2 Tests to establish physical, mechanical and chemical properties of soils

Natural moisture content

Several methods of determining this are given in BS 1377. The most common is to place a weighed sample in an oven and weigh again when dry. The moisture content is always given as the amount of water expressed as a percentage of the *dry* weight of soil. Chemically combined water is not driven off at the temperature at which the soil is considered dry (given in BS 1377). Organic or peaty soils may require a lower temperature.

Soils of greater particle size than silts, when sampled below ground water level, may give very misleading results owing to migration of water during and after sampling. A single moisture content result may also be misleading, as even a relatively homogeneous soil stratum may show a large variation in moisture content, eg London clay may have variations of up to 4 per cent between samples only inches apart. Therefore, the average of several samples is normally obtained. Moisture contents of samples which have been badly sampled, badly packed or have stood for a long time are of little value.

This type of test can be carried out only in the laboratory and the drying time normally causes a day's delay. This is sometimes inconvenient, so quicker and cruder methods, such as drying over a sand tray and primus stove, may be tried. Only certain soils can be tested in this way, and the properties of the soil particles tend to change as a result of localised heating. Therefore, constant attention is required, which increases the cost.

There are chemical methods which depend upon the water in the soil sample producing a gas in reaction with carbide. These methods are not standardised and are inaccurate with clay soils, for which accuracy is particularly important. An in-situ method of measuring moisture content is one associated with methods of measuring the density of soils by means of gamma rays referred to under 'Bulk and dry density' below.

Bulk and dry density

This test determines the weight of the wet soil in situ. When the moisture content is known the weight of the dry soil can be calculated.

For soils which cannot be reached by hand excavation or in trial pits, the only means of obtaining a direct reading of this result is by testing an *undisturbed* sample. In such a situation, no direct reading can be obtained for soils of larger particle size than fine sands. (See Chapter 19)

In undisturbed samples the true value may be altered as a result of the sampling procedure, but this disturbance will be small for most clays in the UK. The result is obtained by weighing a known volume of the soil. If the soil will retain its shape easily on cutting (eg London clay), a known cylindrical volume is cut and weighed. For harder soils and more awkward shapes, the volume of a known weight is obtained by water displacement in a suitable can. When soils absorb water easily the sample must first be coated with paraffin wax. Certain soils give inaccurate results with this method.

When a technician can reach the soil, determination of the density in situ is possible. The methods described in BS 1377 rely on the technique of replacing a known weight of the soil with a standardised material (sand, water or air) which gives the volume of the hole from which the soil is removed. It is possible to cut samples of soil (if the soil is suitable) and make laboratory tests in the same way as for undisturbed samples.

The density of many soils cannot be found by direct reading, as undisturbed samples are unobtainable because the soil is unsuitable and lies too deep for exposure and testing by the replacement method. This applies particularly to sands and gravels, for which there are two other methods.

The first is the standard penetration test (spt), made during the process of soft boring by shell and auger methods (see Chapter 17).

A standardised sample tube, approximately 50mm in diameter by 810mm long with an open-ended cutting edge, is driven into the stratum at the bottom of the bore hole by means of a 63½kg hammer weight with a drop of 760mm. The number of blows required to drive the tube 450mm is counted and the number required for the last 305mm is called the 'blow count.' (these are approximate metric equivalents of standard imperial dimensions). Care is taken to see that the bottom of the bore hole is cleared of disturbed material before the test starts, and that the casing does not extend below the depth of bore hole at which the test begins (see Chapter 17). Such an spt can be carried out every 1 to 2m when boring through a sand or gravel stratum. The procedure is standardised in most countries and provides a reliable indication of the density of the stratum, provided care is taken that the soil is not previously disturbed. Other types of penetration test (see Class 3 Tests—in-situ methods) may also give useful information about density.

Table III *Density of sands related to results of standard penetration tests*

Number of blows	Relative density
0–4	very loose
4–10	loose
10–30	medium
30–50	dense
over 50	very dense

The second method for reading the density in situ relies on the use of gamma rays. The probe may be used on the soil surface or be lowered down a bore hole to measure density at depth. The method has not been widely used but gives a direct result. It can also be used to measure moisture content in situ.

Maximum and minimum densities

The results of the spt can be converted to actual densities when the maximum and minimum densities are found. Such tests can be made only with fine to medium sands and the methods used by Kolbuszewski are often used. To find the maximum density, dry material passing the 4·75mm sieve is rammed into a standard mould. The weight of soil compact gives the dry density directly. The minimum density is found by measuring a known weight of sand which has been poured into a glass cylinder and shaken.

Specific gravity of the soil particles

The result gives the specific gravity—ie the weight relative to the weight of water—of the individual particles of the soil, independently of the bulk density and moisture content of the soil mass. The laboratory test procedure (see BS 1377) varies according to the size of soil particle and is more difficult for fine grained soils, as all air has to be removed. The specific gravity of the soil particles is used to calculate the air content of the soil mass when the bulk density and moisture content are known.

Chemical tests

Although it is possible to obtain a complete chemical analysis of any soil, the routine laboratory tests of this kind are normally confined to:

determination of the sulphate content (this may also be carried out on ground water samples);

determination of the pH value, which measures the acidity of the soil;

determination of the organic content for soils containing peat or similar material.

see BS 1377).

The first two results are used to determine the possibility of action on structural concrete in foundations; and the third to determine whether the soil should be removed because of the large amount of settlement characteristic of organic soils.

Index tests

Index tests are made with fine grained soils only and their results are valuable. Their use overlaps class I tests in that the results are used to classify the fine grained soils. They are grouped by means of the plasticity chart which, with the other tests in classes 1 and 2, makes it possible to estimate the behaviour of the soils under load and, to some extent, their strength. This overlaps tests in classes 3 and 4.

The tests determine the liquid limit, the plastic limit and the shrinkage limit; from which are calculated the plasticity index and the liquidity index. The results are, in fact, the moisture contents at which the nature of the soil changes. The liquid limit is the moisture content of the soil at which the soil will begin to 'flow' like a liquid under certain conditions. The plastic limit is the moisture content at which moulding of the soil ceases to be possible under certain conditions and the soil becomes fragmented. The shrinkage limit is the highest moisture content at which the volume of the soil remains constant when dried. Above this limit the soil will shrink when dried. (See BS 1377.)

Class 3 Tests to establish the strength of soils

Laboratory methods

Although these tests are made by similar methods throughout this country, standard procedures have not been laid down in all cases. Therefore the more common tests only are described here.

Laboratory testing of compression and shearing strength requires a sample of a regular shape to allow accurate calculation of the stress at failure. The load on the sample may be measured hydraulically, with proving rings, or with calibrated springs. The simplest form of test is the unconfined compression test, in which a sample of soil is crushed, the maximum load being measured. The sample is held only at the ends during the test and, therefore, only clay can be tested by this means, it being the only soil which is capable of holding together after being cut.

A simple portable apparatus, which measures the load applied to the sample with calibrated springs, has been developed by BRS. The test is made with undisturbed samples (see Chapter 17), and can be carried out on site. but is of limited value except as a check on approximate strength.

The most commonly used compression test is the triaxial compression test which, in its various forms, can be used for all soils. The sample is cut from bore hole samples or is made up in the laboratory. It is subjected to hydrostatic pressure and then tested by a vertical compression load. Sand and gravel samples are cast in a mould and transferred to the testing cell under a negative pressure which maintains their shape. Many variations of the test are possible for different problems. The rate of load application, rate of strain and confining pressure can be varied.

The shear strength can be calculated from the

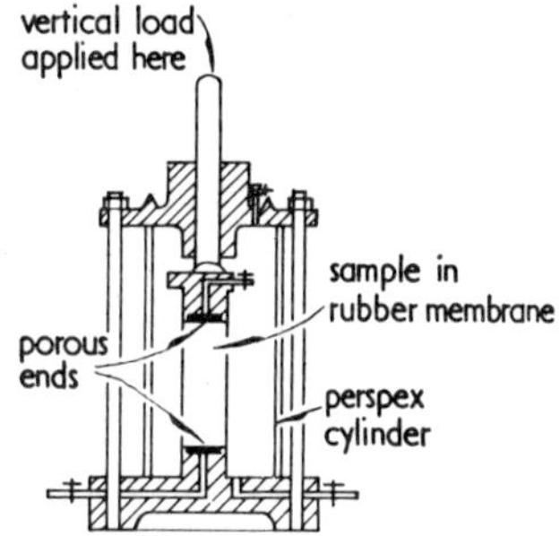

2 *Section of triaxial compression apparatus*

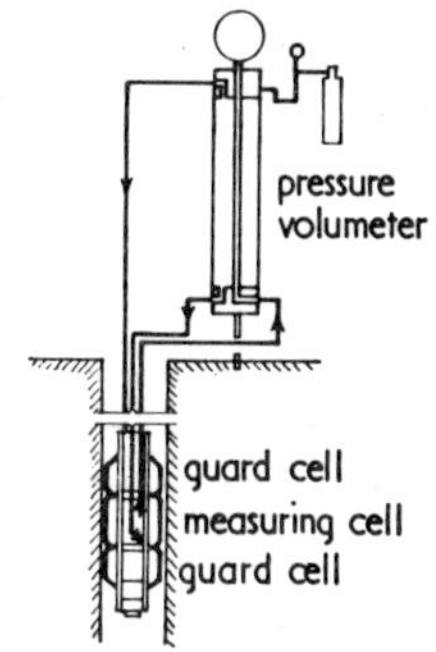

3 *Section of the Menard pressiometer in use*

results of the triaxial test: it is the most important factor used in calculating the bearing capacity. The direct reading of this strength is difficult, and for this reason the results of the compression test are normally used. However, direct shear strength measurements may be made for sands and gravels in a shear box.

In strength testing where samples cannot be obtained undisturbed from the bore holes, the sample is made up at the density and moisture content at which it exists in the ground, using test results.

Some alluvial clays are 'sensitive' ie there is an appreciable reduction in shear strength when they are disturbed, for example during sampling or in the course of construction. It may be necessary to obtain their undisturbed strength by an in-situ method, but it may also be desirable to measure the reduced strength of the remoulded samples.

In-situ methods

In site testing, it is possible either to measure the strength experimentally, or to measure physically the load which the soil will bear. The actual measurement of strength in situ may be difficult because complicated apparatus is often required. Such tests if carried out at various depths without pre-boring are often called 'deep soundings' and are limited to soils which can be easily reached; to be able to reach to any appreciable depth normally means that the soil is weak. In the case of very soft silts and clays, the use of in-situ tests is very often necessary because sampling is impossible. For such soils, the vane test is used.

In this test, a four-bladed vane, about 50 to 100mm wide × 75 to 150mm deep, is rotated at the bottom of the bore hole by rods. The torque required to rotate the vane is measured and used to calculate the shear strength of the soil. In soft ground the vane can be pushed down without a bore and the shear strength measured at various depths. In this case the vane rod may be enclosed in a casing tube to eliminate the effect of soil friction on the rod. There is a laboratory form of this test in which the vane is rotated in a sample, but it is not much used in this country.

Another instrument which measures the strength directly is the Menard Pressiometer **3**. This requires bore holes to measure the strength at depth but they are only 2in in diameter which is smaller than a normal bore hole. The strength of the soil is given from the pressure required to inflate a water-filled balloon which fits into the bore hole. The test may be used at any reasonable depth but it is less reliable in clays because of their slow drainage properties.

The most frequently used in-situ test is the dynamic standard penetration test described under class 2, which is used to determine the relative density of a soil. The strength of the soil is obtained from graphs which relate the results of the spt to allowable bearing capacity, based on tests made throughout the world. The spt is unreliable in clayey soils. When the penetration test is used to give a continuous reading it is called the 'dynamic cone penetration test' or 'static cone penetration test'. The latter is very widely used in the Netherlands to predict the bearing capacity of piles. Instead of being hammered into the ground at intervals, the cone is driven until it stops by hydraulic or mechanical pressure, which is continuously read. The cone is driven down until the load available is not enough to drive it further. The driving rods are hollow, the cone being driven ahead of the rods by an inner piston rod sliding inside the outer rods. This enables the resistance caused by friction on the rods to be determined separately from total resistance, which is of great advantage when dealing with pile calculations.

This test is used mainly in soft silts and sands, and is of great advantage in esturine conditions, as it can be made from a boat. The graph of continuous cone resistance with increased depth is used to calculate probable settlement and the allowable load. In the case of piles, the test is obviously a scale model test of an actual pile.

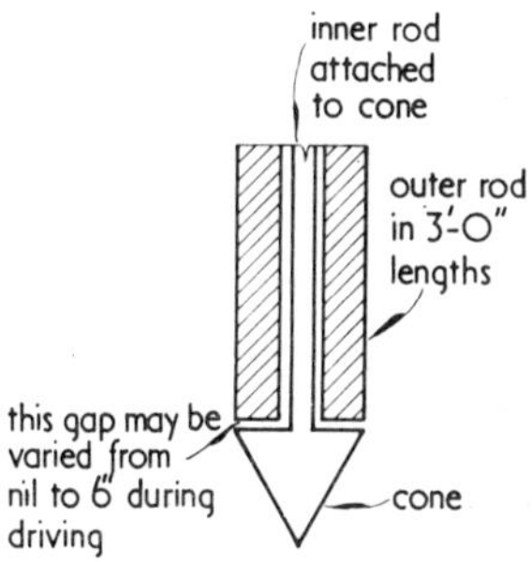

4 *Section of cone used in static penetration test*

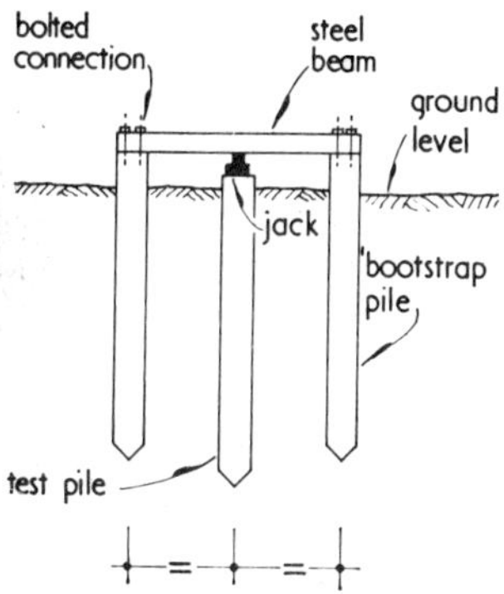

6 *'Bootstrap' method of jacking a test pile into ground*

Class 4 Tests to establish the behaviour of soil under a specific type of loading or stress, or in the course of time

Laboratory methods

The most common laboratory test is the consolidation test which is used to predict the settlement of buildings when built on clay. It can be applied only to clays which cannot be tested for loading in situ. A small sample of the soil is cut from undisturbed samples in a cutter ring and placed in a consolidation press (oedometer) which loads the sample increments up to about 1700kN/m^2. Each increment of load decreases the thickness, and the amount and rate of decrease are measured. The results enable the total settlement of a structure to be predicted and the rate of settlement to be estimated. Swelling

U

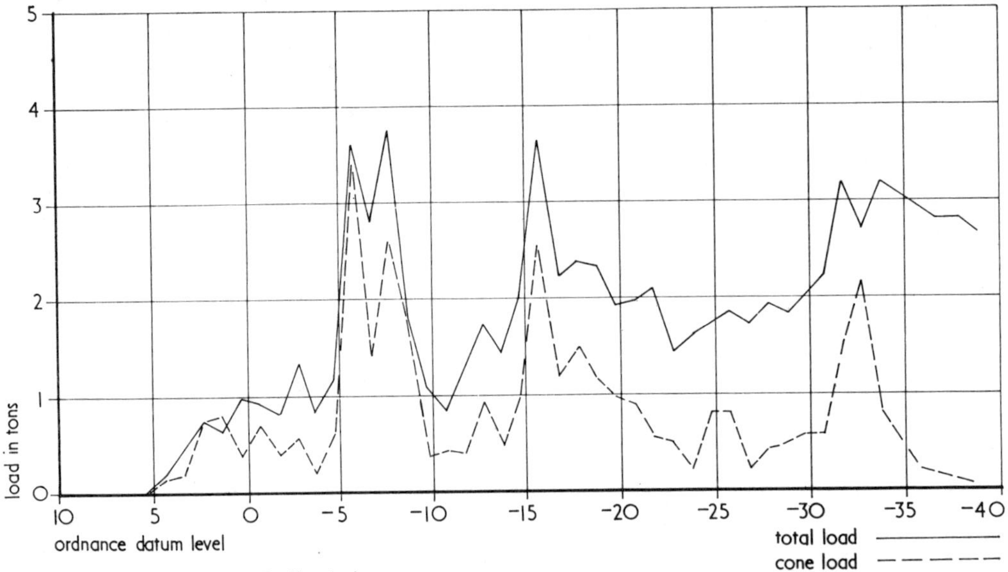

5 *Typical result of cone penetration test*

or heave of clay under decrease in load (eg in excavations) can also be determined. Calculations take the average of several tests.

Consolidation properties of clay can also be obtained in a test using the same cell as the triaxial compression test. The sample is prepared as for a triaxial test and the required pressure applied in increments to the water in the cell. The resultant change in the volume is measured from the amount of water squeezed out.

In-situ methods

For sands and gravels, the settlement is calculated from the density or the penetration test or obtained from a loading test, which will also give the bearing capacity.

In mixtures of soils eg fills and soils made up of thin bands of different types, the plate bearing test may be the only method. This can, however, test the soil only to a depth of about 1·5 times the plate diameter, and the results need very careful interpretation. A round or square plate is loaded until failure of the soil results. The load, which may be 50 tonnes or more, is applied either by dead weight or by jacking downwards off kentledge. Both methods require large amounts of heavy materials. Plates 600, 900 or 1200mm in diameter are commonly used; a larger plate stresses a greater depth of soil. Tests using two sizes of plate yield a better result. The downward deflection of the plate at more than one point is read to allow for possible tilting. Load increments are applied only when settlement after the previous increment has ceased and the test, therefore, may last several days. Long term settlements of clay cannot normally be determined by loading tests.

Allied to this test is a full scale pile test. The theoretical calculations of pile capacity are notoriously inaccurate and a pile test, although expensive and lengthy, is the best method of assessing bearing capacity. Load is applied to the head of a pile, driven to the required depth. Again, the weights or kentledge necessary are very considerable.

The necessary reaction is often supplied by the 'bootstrap' method, whereby the pile is jacked into the ground, using as a reaction steel girders connected to two, three or four piles surrounding the test pile.

In an improvement to this test, the pile is driven down continuously at a constant rate without waiting for movement to cease at each load increment. The method gives very accurate results for the bearing capacity but is not a good check on working load settlement. The problems involved in providing sufficient weight are increased by the need to provide a 'pacer' to keep the rate of penetration constant, without which the result may be useless.

Tests for roadworks and earthworks

The California bearing ratio test (see AJ Information Sheet 1012, AJ 18.10.65) which is based on experience rather than theory, compares a soil with an 'ideal' material for use as a road base. A plunger is driven into a sample in a container. The load required to cause a standard penetration is measured and compared as a percentage with the load required in the ideal material. The rate of penetration is kept constant. The sample may be undisturbed or, usually, it is made up to a known density. Sometimes it is saturated. The test may also be used in

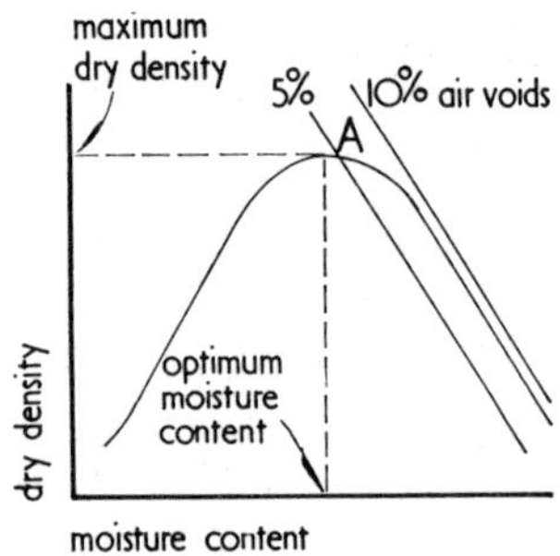

7 *Curve of dry density against moisture content for compaction test*

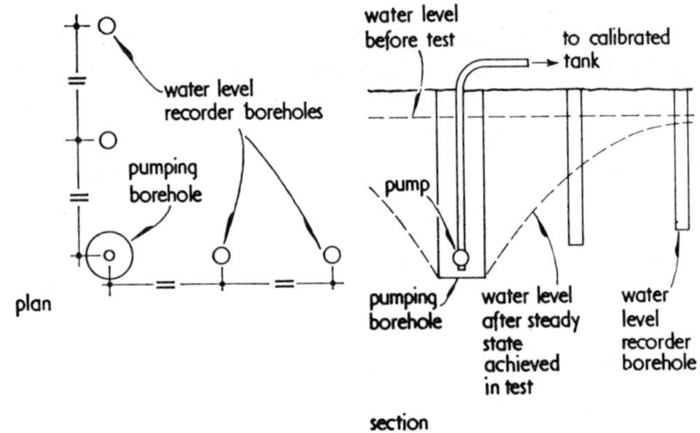

8 *Pumping test: plan and section of pumping bore hole and recorder bore holes*

situ directly into the soil or road base.

The compaction test gives the maximum density that can be obtained with a particular soil and the moisture content at which this will occur, ie the optimum moisture content. The soil is dried and differing amounts of water are added to several samples. These are compacted into a standard mould and the results are used to plot a curve of dry density against moisture content so that the dry density to be achieved in earthworks compaction can be specified. Also plotted on the graph are two lines of constant air voids; as with earthworks, a maximum figure of air voids may be specified instead of a density.

Drainage

For drainage, permeability is the important factor. Clay has low, and sands and gravels high, permeability.

Two laboratory tests are used to measure permeability: the constant head permeameter test; and the falling head permeameter test. The first is for sands, gravels etc, and the second for fine grained soils, (a direct test is not practicable in the case of comparatively impermeable clay, but the coefficient of permeability can be calculated from the results of a consolidation test). In the first the sample is made up to a known density, while in the second undisturbed samples are used. In both tests water is passed through the sample and the hydraulic gradient and quantity are measured. In the constant head test, the head of water causing the flow is kept constant while in the falling head test, the head is allowed to fall. This test is on a small sample only and ground conditions are always more complicated than laboratory simulation. Greater accuracy is achieved with a full scale pumping test in situ.

In the pumping test a bore hole is sunk to a depth below the stratum whose permeability is to be measured. Further holes are sunk at points around the main bore hole in which the water levels are recorded before, during, and after the test. The bore hole is pumped continuously, the quantity of water being measured in a tank with a calibrated weir. Water levels are recorded until all levels are stationary. This test is expensive, as the steady state may take several days to reach. From it is calculated the actual permeability of the ground, used to design de-watering systems, retaining walls and coffer dams in ground with a high water table. The results can also be used to design soakaways and so on.

Class 5 Tests for stabilised soils

General

When soil is used as a constructional material it is usually moved to a new position and it is necessary to control its quality in the same way as other materials.

The common properties which must be checked by the relevant tests are: moisture content; dry density; air voids; California bearing ratio and grading.

Additives

Sometimes the soil may have additives to increase its strength eg cement, lime or bituminous materials. Their properties, such as density, strength, water absorption, can be tested in the laboratory (see BS 1924).

Pore pressure tests

Where the soil is used as the main structural material, eg in an earth dam, the usual properties are measured, together with pore pressures inside the soil mass resulting from compaction. The pore pressure is that exerted on the individual soil grains by the free water in the voids between particles. These are measured by hydraulic or electric gauges buried in the soil mass as construction proceeds.

Change of properties

Soil can be used as a construction material in its original position with its properties changed by chemical injection, cement grouting, electro-osmosis, freezing, or vibroflotation, the choice of method depending on the requirements and on the existing soil properties. The tests used to discover the properties before and after change will be those discussed in classes 1 to 4.

Chapter 19
Soil investigation: Documents necessary for a report

Architects frequently need to obtain details of the type of soil likely to be encountered on a site and therefore have to commission an engineer or specialist firm to undertake a soil investigation.
This chapter discusses the ways in which the investigation work can be allocated; mentions the methods, types of sample etc, and how they should be specified for tests on and off the site; and describes the form in which the report should be requested.

SOIL INVESTIGATION REPORT: GENERAL

Documents

In order to obtain competitive tenders and to ensure that the work is adequately controlled, it is normally very desirable that the requirements for a soil investigation and report should be fully documented before any work is carried out. Such documents can be drawn up on the basis of quantities in a similar way to other work in the building and construction industry. The main requirements are that the person who prepares the documents has knowledge of the methods and processes, their application and the use of the results; and that the general outline of the proposed structure is known. An economical soil investigation must assess the feasibility of a particular proposal. A general investigation with no proposal in mind is likely to be either too large or too small in relation to whatever scheme is finally evolved. The documents must say how the soil investigation is to be done; and they must define the extent of the investigation. In view of the specialised nature of the work it is often advisable to say who is to be responsible for the different parts.

The work may be divided into the following stages:
a work on the site;
b work off the site, in a laboratory;
c the combined results and conclusions of a and b: the report and design application.

Who is to do the work?

There are four possible methods of organising the work (see Table 1).

Many contractors will carry out all three stages of the work. However, if anything suffers in this method it is the report and conclusions. The demarcation between advice and detailed design is difficult to draw and even more difficult to cost. Therefore, the report may tend to evade difficulties, or even miss the point altogether.

The work done off site in a laboratory should be under the supervision of an engineer familiar with the design problems of the particular investigation; and the results and conclusions must be the responsibility of a competent engineer. Some consulting engineers have laboratories under their own control and do *b* and *c* as preliminary work to design. In such circumstances, the site work is done by specialist contractors which suits the present organisation of the industry. Provided that the party carrying out *c* is connected with the design process (ie is the consulting engineer for the structure or a specialist consulting engineer) then the fullest use is made of the information obtained on site and in the laboratory.

It can be seen, therefore, that documents may be needed for any one or all of *a*, *b* or *c*. But whatever is needed, each compartment of the work must be fully detailed to allow for comparison of work, if required, and to make sure that the amount of work being done is applicable to the problem to be solved, and also to change the requirements if this becomes necessary as the work reveals hidden snags.

Table 1 Ways of allocating work for a soil investigation

Method	**Party to carry out each stage of work**		
	Work on site	**Work in laboratory**	**Report and design application**
1	Specialist contractor	Specialist contractor	Specialist contractor
2	Specialist contractor	Specialist contractor	Consulting engineer
3	Specialist contractor	Consulting same as engineer	Consulting engineer
4	Specialist contractor	Consulting laboratory or engineer	Consulting engineer

SITE WORK

Scope

This covers:

(*a*) boring methods and quantities,

(*b*) sampling,

(*c*) tests carried out in-situ or otherwise on site,

(*d*) specialist site work,

(see Chapters 17 and 18.)

In general, site work is best carried out under the *General conditions of contract for use in connection with works of civil engineering construction* (obtainable from the Institution of Civil Engineers and the

Federation of Civil Engineering Contractors). If, however, it is desired that separate conditions of contract be written into the documents for a particular investigation, this may be done provided care is taken that they are adequate having regard to all the circumstances.

Boring methods and quantities

The methods used to reveal the soil or rock strata on a site are very important. They must provide the quickest and cheapest penetration of the strata, revealing the largest cross-section with the least possible alteration to the strata. The choice is between hand methods and machine.

Hand methods

These are suitable to depths of about 3 to 4m. They are particularly useful for exposing existing foundations. They can reveal more of the strata—in trial pits—than machine methods. Hand methods fail on grounds of expense in waterlogged ground, rock strata, or strata with very large stones or boulders.

Machine methods

These are of two general types: those designed to penetrate 'soft' strata and those designed to penetrate rock. The latter are dearer than the first, but will normally do both jobs if necessary (although soil sampling will be difficult or impossible in 'soft' strata), whereas the first will not do the second. The dividing line between 'soft' and rock strata depends to some extent upon the drill and the driller. Generally speaking, hard marl or a mudstone is the hardest 'soft' stratum economical to drill by 'soft' boring methods.

Statement of method to be used

The documents must say which type and size of boring is to be used. If more than one type is required these must be stated separately. If necessary, unsuitable methods may be specifically forbidden. If it is necessary to change methods during the site work, the reasons for the change must be given.

Amount to be done

The documents must make clear that the contractor is responsible for bringing the equipment and labour for each method required to the site and away again. Items must be included for setting up the necessary equipment at each position where boring or digging takes place. These items should include for clearing up each position and leaving it in a satisfactory condition.

The actual penetration of the strata should be

itemised with the total linear amount of boring followed by the depth of the holes eg:
50m between 9m–12m below surface
25m between 12m–15m below surface
in a manner similar to trench excavation. The depth increment is normally 3m, but may be varied. In view of the uncertain nature of site investigation work much use of provisional items is advisable in case the need for more or less boring becomes obvious. As a precaution, items for penetration deeper than expected should be included. Separate items should be included for different methods, which may mean two or more complete sets of boring items for each job. These items should provide for the recording of boring ie the bore hole logs.
If necessary, the items similar to the boring items should be included for backfilling the completed bore. Any special backfilling must be itemised separately.

Difficult conditions
In cases of conditions peculiar to a particular site, items should be inserted to cover these so that the contractor is aware of them. Possibilities are:
restricted access,
restricted headroom,
underground services,
watching, lighting or protection,
staging at bore hole positions,
standing water at bore hole positions.
It is important that all site work be related to a reduced level of reference of some kind. The documents may include an item to provide for the levelling of the bore hole or trial pit position.

Sampling
The methods for taking samples from the stratum penetrated follow from the boring methods. 'Soft' boring sampling methods cannot be used in rock boring, but can often be used in trial pits. Water samples can be taken in any type of bore. (See Chapter 17.)

Methods to be used
The documents must say what type, size and depth of samples are wanted. They must also say how they are to be taken, protected and stored and where they are to be sent for testing and examination. If long term retention of samples is required, this should be specified for each sample type. Samples can be kept until the construction contract is begun to allow for use during the tender period' Sometimes samples are kept for comparison with conditions found during construction. Normally

samples are disposed of four to six weeks after the completion of the report.

Amount to be taken

The expected quantity of each individual type of sample must be stated. Provisional quantities of each type of sample should be added to allow for contingencies.

In certain cases, it may be convenient to include under this heading examination of samples by a geological specialist. Such work is often done on the site rather than in the laboratory.

The documents should provide for recording all samples in the bore hole or trial pit logs.

Tests carried out on site

These may be done in-situ or may be laboratory type tests on samples. The former are divided into those carried out in the bore holes and those done separately. These last may be done at the same time as the boring or they may be of such size and complexity as to require separate arrangements eg full pile tests. In such cases, a separate specialist contractor may be used, or the contractor for the boring may use such a specialist as a subcontractor.

Method to be used for in-situ tests

The procedure for these tests varies more than normal laboratory testing, according to the site, the design problem and the particular engineer's approach. Careful specification is important, having regard to duration of tests, the size of test equipment, and the accuracy of measurement of readings. It may be necessary to provide for the continuous reading of results over long periods. Tests may be limited by the equipment available and sometimes the construction of special equipment may be necessary.

Specifications for tests should be written precisely and in collaboration with the design engineers. Standard procedure rarely applies, but for standard penetration tests and constant rate of penetration pile test reference should be made to the prime source of the standard procedure. If properly specified and carried out, in-situ tests are of more value than any other form of investigation.

Amount to be done

The tests should be billed by item. They will in most cases be costly and, therefore, a close and continuing check on the numbers should be made during the site work. Provisional items will again be useful in case unforeseen difficulties need resolving. The items should include the reporting of results only. The interpretation will be part of the report.

Specialist site work

Under this heading comes:
geophysical work,
corrosive action survey,
special geological investigation,
mining subsidence survey.

These are included for the sake of completeness, but will occur rarely (except perhaps for mining subsidence problems) outside heavy civil engineering or marine works. The architect will be guided by the circumstances, but most probably the whole of the problem will be investigated and reported on by an expert consultant. If not, any site work can be included in the site testing as a special item, if necessary, by a nominated subcontractor.

It should be pointed out that geophysical methods (electrical, seismic, magnetic) can be of value in delineating changes in markedly different strata and discontinuities such as old mine workings, but the very greatest expertise is required if the information obtained is not to be of doubtful value or even positively misleading. Results should normally be confirmed by conventional boreholes, and perhaps geophysical methods are of greatest use in carrying out a rapid check of the consistency of certain strata between borings and in pin-pointing doubtful areas for further investigation.

WORK OFF THE SITE

Laboratory testing

This is almost always the testing of samples obtained at the site and is done in the laboratory of the firm which does the site work, in an independent laboratory, or in the laboratory of the engineer concerned with the design. The laboratory testing programme should be specifically designed to solve the actual problems posed by the site. This is best done by the design engineer. In this way unnecessary tests—done under the heading 'normal tests for that type of soil'—can be avoided.

Procedures for laboratory tests are more standardised than those for site tests. The two sources of standards are British Standards and a few textbooks on the subject. This does not mean that non-standard tests cannot be used when required.

Method of testing

When the testing is to be done by a laboratory not under the control of the design engineer, the docu-

ments controlling the work must cover the following points:
the name of the test,
the procedure of the test be reference to a standard if possible, and if not by detailing the procedure,
how the result is to be presented,
what information is required in the result,
what units the result is to be in.

If possible the laboratory work should be controlled so that any deviation from non-standard procedure or any unusual occurrence during testing will be taken into account when the results are presented. 'Routine testing' is of little use when difficulties arise.

The work to be done should be itemised test by test, with provisional items to cover the possible need for more tests.

THE REPORT

This should be called for in two parts:
(*a*) what has been done,
(*b*) the conclusions and recommendations.

What has been done

This part can be written by the persons carrying out the work. It should repeat, in a readable form, almost all the information that has gone into preparing the documents controlling the work.

1 The site work report for boring and/or trial pits and sampling should state:
number, method used, size, position and depth,
strata penetrated,
water conditions found, (strikes and equilibrium conditions).
position, types and number of samples taken,
date and length of work,
any unusual features.
Nearly all this can be put in the bore hole logs and/or site plan.

2 The site work report for in-situ tests should state:
procedure or reference to standard procedure,
numbers and positions,
results.

3 The laboratory tests report should state:
name of a standard test,
procedure for a non-standard test,
results in a clear and readable manner.

4 There should also be a general summary of the geology and topography of the site.

Conclusions and recommendations

No sensible conclusions and recommendations for a site investigation report are possible unless the person writing the report knows the following (at least approximately):

1 height of proposed structure,
2 plan area of proposed structure,
3 loads and positions of loads at foundation level,
4 approximate allowable differential movements in proposed structure,
5 general requirements for any soil retaining structures, roads or earthworks.

Moreover it is necessary that he should be aware of any topographical features of the site which could affect his recommendations, particularly in regard to possible ground water movements. These include vegetation and drainage, chances of flooding, tidal conditions, details of properties, installations or constructional work in progress (especially pumping) on or adjoining the site, and 'swallow holes' or 'sink holes' which could indicate underground cavitation. The condition of adjacent structures may be of relevance.

It is also helpful if certain types of foundation are known to be impermissible owing to special factors, eg cost. Any report written for a site investigation in which these factors are unknown will either have to be so vague as to be useless, or have to cover all possibilities.

As previously stated, a better, more useful and more economic investigation is obtained by having the investigation controlled by an engineer who knows these factors beforehand. It is obvious that the best system is for this person to write the report also.

The recommendations should give:

1 preferred types of foundations in order of precedence from an engineering and economic standpoint,
2 preferred method of construction of these foundations,
3 special precautions to be taken before, during, or after construction,
4 special precautions to be taken during construction of the superstructure eg symmetrical loading of foundations,
5 special points such as drainage, mining subsidence and so on,
6 design and construction recommendations for any soil retaining structures, roads or earthworks
7 prediction of any settlements or other soil movements likely to occur during or after construction.

If the proposed project is not feasible on engineering and/or economic grounds this must be stated.

The recommendations cannot give the detailed design and size of the recommended foundation type. This is part of the design process, because the size, depth etc, of a foundation depends upon not only the load but also the point of application and size of application of the load on the foundation. The report should refer to those matters which are still doubtful even after the investigation, and which may cause changes of design before or during construction as more knowlege is obtained. When the documents for a site investigation call for the writing of a report, these documents should therefore list all the known factors concerning the proposed structure and call for the report to give the necessary recommendations listed above.

Bibliography

General

KELLY. Kelly's manufacturers and merchants directory 1971–72, vol. 1, UK and Republic of Ireland, vol. 2, International and separate Trades and Services Index. Kingston upon Thames, 1971, Kelly's Directories, 85th edition, £8 per set. /(Abd)/

STAMP, D. L. Britain's structure and scenery. London, 1967, Collins, 6th edition, £2·25 /00(E4)/

TRUEMAN, A. E. Geology and scenery in England and Wales. Harmondsworth, 1961, Penguin Books, £1·25. /00(E4)/

Planning

DEPARTMENT OF THE ENVIRONMENT. Welsh Office. Sunlight and Daylight. Planning criteria and design of buildings. London, 1971, HMSO, 40p /(N7)/
(Sunlight and Daylight Indicators are published separately. London, 1971, HMSO, £2·50 per set.)

HEAP, D. Outline of planning law. London, 1969, Sweet and Maxwell, 5th edition, £2·75, Paperback edition £1·75. /05(Aj)/

HEAP, D. and BURKE, J. (editors). Encyclopaedia of the law of town and country planning. London, 1959, Sweet and Maxwell, Reprinted May 1972, £16·80 /05(Aj)/

(Three loose leaf volumes, so that new material issued periodically may be inserted and out-of-date sheets removed)

Site surveying

BOWYER, J. T. Guide to domestic building surveys. London, 1971, Architectural Press. £2·50. /(A3s)/

BRITISH STANDARDS INSTITUTION: Site investigations. BS CP 2001: 1957. London, The Institution, £1·90. /(11)(A3s)/

CLARK, D. Plane and geodetic surveying for engineers, vol. 1. London, 1969, Constable, 6th edition revised by J. E. Jackson. 1972, 6th edition metric. £2·25. /(A3s)/

PRIME, C. T. and DEACOCK, R. J. Trees and shrubs: their identification in summer or winter. Cambridge, 1971, W. Heffer and Sons, 6th edition revised and enlarged. 65p. /Yx/

REEKIE, R. F. Draughtsmanship. London, 1969, Edward Arnold, 2nd edition. £1·40 /(A3t)/

ROYAL INSTITUTE OF BRITISH ARCHITECTS: Handbook of architectural practice and management. London, 1967, The Institute, 2nd edition, £10. (under revision) /(A1)/

STOKOE, W. J. The observer's book of trees. London, 1961, Frederick Warne, 2nd edition, 45p /Yx1/

TOMALIN, G. Precision site surveying and setting out. London, 1964, Adam Hilger, £4 /(A3s)/
WHYTE, W. S. Basic metric surveying. London, 1969, Butterworth, £3·20, Paperback edition £2·25. /(A3s)/

Noise

ARCHITECT'S JOURNAL. AJ Handbook of Building Environment, Section 5, Sound. Arthur Aldersley Williams. AJ 22.1.69–12.2.69. London, Architectural Press
BRITISH STANDARDS INSTITUTION. Sound insulation and noise reduction. BS CP 3: Chapter 111. 1960, London, The Institution, £1·90. /(M)/
HARRIS, C. M. (editor) Handbook of noise control. 1957, McGraw-Hill, £12·35. /(M5)/
NATIONAL PHYSICAL LABORATORY. Notes on Applied Science 10, Noise measurement techniques. London, 1961, HMSO. Out of print (Series now discontinued)
PARKIN, P. H. and HUMPHREYS, H. R. Acoustics, noise and buildings. London, 1969, Faber and Faber, 2nd edition, £5, Paperback edition £1·75. /(M)/

Climate

ARONIN, J. E. Climate and architecture. New York, 1953, Reinhold /(E7)/ out of print
Popular approach to the effects of sun, wind, temperature and precipitation on design. Mainly but not exclusively

American experience

BUILDING RESEARCH STATION Digest 127. An index of exposure to driving rain. 1971, HMSO 5p. /(E7)/

Contains maps showing driving rain conditions over the British Isles and suggests the use of a numerical driving rain index

SHELLARD, H. C. Microclimate and housing. AJ 1965, January 6, p. 21–26, January 13, p. 81–84 /(E7)/

AJ technical studies which discuss topographical effects, factors concerning orientation, and how houses themselves affect microclimate

LAMB, H. H. The English climate. London, 1964, English Universities Press £1·05 /(E7)/

Excellent authoritative introduction for the layman

METEROLOGICAL OFFICE MO 488. Climatological atlas of the British Isles. 1952, HMSO out of print /(E7)/

The only comprehensive collection of maps representing climatological data in the British Isles

METEROROLOGICAL OFFICE Climatological Memorandum 43 (monthly and annual maps of mean daily maximum and mean daily minimum temperature, and of average summer and winter temperature over Great Britain and Northern Ireland, 1931–1960) /(E7)/

Smaller version of the *Climatological atlas* and dealing only with temperature

GEIGER, R. The climate near the ground. 1965, Harvard University Press 3rd edition, £7·25. /(E7)/

Standard comprehensive work on the subject but of only specialist interest to the architect. Very extensive bibliography

BUILDING RESEARCH STATION Current paper 5/72. Survey of meteorological information for architecture and building by R. E. Lacey. 1972, HMSO /(E7)/

CHANDLER, T. J. The climate of London. London, 1965, Hutchinson £3·50. /(E7)/

Thorough analysis of London's climate and the variations resulting from changes in the city's morphology

INSTITUTION OF HEATING AND VENTILATING ENGINEERS Guide. Book A. London, 1970, The Institution

Essential information for all aspects of heating and ventilating design

Soil investigation

AKROYD, T. N. W. Laboratory testing in soil engineering. Bracknell, Berks. 1970, Soil Mechanics Ltd., 1st edition with revisions. £3. /(L4)/

ARCHITECT'S JOURNAL. AJ Handbook of Building Structure, Section 4, Foundations and retaining walls. Hodgkinson, Allan (consultant editor). Published Spring 1972. London, Architectural Press. /(2-)/

BRADE-BIRKS, S. G. Teach yourself good soil. London, 1959, English Universities Press Ltd., 2nd revised edition. 52½p. /(L4)/

BRITISH STANDARDS INSTITUTION. Test sieves. BS 410: 1969, Metric edition. London, The Institution £1·00 /(L4)/

BRITISH STANDARDS INSTITUTION. Methods of testing soils for civil engineering purposes. BS 1377: 1967, London, The Institution. £2·75. /(L4)(Aq)/

BRITISH STANDARDS INSTITUTION. Methods of test for stabilized soils. BS 1924: 1967, London, The Institution, £2·75. /(L4)(Aq)/

BUILDING RESEARCH ESTABLISHMENT (formerly BRS)
Digest 63 Soils and foundations: 1. 1965, Metric revision 1970. London, HMSO, 5p
Digest 64 Soils and foundations: 2. 1965, revision 1969. London, HMSO, 5p
Digest 67 Soils and foundations: 3. 1966, revision 1970. London, HMSO, 5p. /(16)(L4)/

HOUGH, B. K. Basic soils engineering. New York, 1969, The Ronald Press Company, 2nd edition, $13·50. /(L4)/

TERZAGHI, K. and PECK, R. B. Soil mechanics in engineering practice. New York and London, 1968, John Wiley, 2nd edition. £7·50, Paperback edition £3·95. /(L4)/

TOMLINSON, M. J. Foundation design and construction. London, 1969, Pitman, 2nd edition. £6·50 /(16)/